THE QUR'AN AND QUANTUM PHYSICS

HEART & SCIENCE

WRITTEN BY SEBAHAT MALAK

Sebahat Malak

Copyright © 2018 Sebahat Malak
All rights reserved.
ISBN: 9781731545305

PREFACE

Modern day human beings have come to the realization that science and religion are not separate or distant concepts. The number of those who do not take science as simply a means of material development increases day by day. Material development can only be of benefit if it is implemented into and intertwined with spirituality. When these two types of formations are handled together, the door to success will open up. This book is the product of the perspective, which demonstrates the importance of creating and executing systems in accordance with human nature, providing the reader with the opportunity to examine a wide range of subjects from quantum physics to art, from divine thought to knowledge.

Step-by-step descriptions from the works of many famous physicists such as Heisenberg, Hawking, Einstein, and Bohr allow us to see how science and religion intersect through the verses from the

Qur'an. Setting aside the pervasive material bias of science will reveal a surprisingly clear consensus between science and religion.

Are books really a work of fiction? Or is it the desire of the author of the book to share with the reader the truths regarding *multi-worlds* of her just like in quantum physics? Isn't her life a glimpse of the worlds that complement her suffering, happiness, love, goals in life, and knowledge of these pieces? How accurate is it to name this work a fiction or imagination? Is fiction, like quantum theory, really just a collection of some people's imaginary or existing worlds?

Everything derives from a glance of Allah, in other words, just one of His words. All the accumulation of knowledge leads us to this very word. Conveying any knowledge to Allah can only be achieved with the permission and will He has above His servants. Knowledge transforms into concepts. That is, knowledge is knowledge and it is the revelation of a morsel of Allah's words. This situation can be expressed by an example: Is it correct to say that the atom is equal to Fatiha, and if so what makes it correct? As is expressed throughout the Qur'an, no one except Allah, not even the prophets, can *"bring a verse down,"* and it is He *who taught by the pen - taught man what he did not know (Al-Alaq 96:4-5)*. Considering this situation, we understand that it is not possible for human beings to bring about what does not exist. The subject can only combine what exists and bring into existence through Allah's command. For this reason, Fatiha's equality with an atomic equation was perhaps not considered previously because the time for an explanation had not yet arrived. Now, if it can be found in the thoughts of a person, then Allah must have commanded it, and hence, it is possible to justify the equation since this is also explained in the process.

Table of Contents:

THE QUR'AN AND QUANTUM PHYSICS ---------14

HEART & SCIENCE-------------------------------------14

1 INTRODUCTION --17
1.1 LIVING UP TO THE FATIHA ----------------------------------17
1.1.1 THE FORMULA OF THE UNIVERSE ------------------17
1.1.2 LIVING UP TO THE FATIHA ------------------------------18
1.1.3 THE SPATIAL SIGNIFICANCE OF FERTILE LANDS ---21
1.1.4 HUMANS AND TIME---------------------------------------25
1.2 CONFUTING SUPERSTITIONS----------------------------25
1.2.1 MEANING IN THE KNOWLEDGE OF THE APOCALYPSE---25
1.2.2 THE PROPHET MOSES ----------------------------------27
1.2.3 THE VIRTUE OF LOVE-----------------------------------29
1.2.4 CONFUTING SUPERSTITIONS -------------------------32
1.2.5 GOODNESS AND EVIL-----------------------------------34
1.2.6 THE FEAR OF ALLAH-------------------------------------37

2 SHAPING TIME ACCORDING TO SPACE AND KNOWLEDGE --39
2.1 OVER EVERY POSSESSOR OF KNOWLEDGE IS ONE MORE KNOWING --39
2.1.1 HERMENEUTICS (INTERPRETATION) ---------------39
2.1.2 DETERMINISM --40
2.1.3 REVELATION AND THE CALL--------------------------42
2.1.4 SHUAIB: THE PROPHET OF JUSTICE IN THE WORKPLACE --45

2.1.5 THE PROPHET SALEH AND THE SHE-CAMEL REPRESENTING CREATION --------------------------------------- 47
2.1.6 ARE THEY MORE THAN NECESSARY, OR ARE THEY DEFICIENT? --- 48
2.1.7 HORN-TRUTH-THE DAY OF JUDGMENT ----------- 49
2.1.8 WITNESSING --- 51
2.1.9 THE NOBLE ALLAH AND THE UNKNOWN --------- 54
2.1.10 OVER EVERY POSSESSOR OF KNOWLEDGE IS ONE MORE KNOWING -- 56
2.1.11 THE EVIL IN UNNECESSARY REQUESTS -------- 58
2.2 THE SURAH AL-KAHF AND THE WINDOWS OF KNOWLEDGE --- 62
2.2.1 THE SURAH AL-KAHF AND THE WINDOWS OF KNOWLEDGE --- 62
2.2.2 THE OWNERS OF THE GARDEN ----------------------- 63
2.2.3 GRACE -- 65
2.2.4 THE GRACE OF FRIDAY ------------------------------- 66
2.2.5 THE KEYS TO THE INVISIBLE REALM -------------- 68
2.2.6 VERSES AND SUSTENANCE ------------------------- 71
2.2.7 THE YOUTH OF ESHAB-I KEHF ----------------------- 74
2.2.8. THE STORY OF THE PROPHET MOSES AND THE PROPHET KHIDR AND THAT ALLAH IS CLOSER TO HIS SERVANTS THAN THEIR JUGULAR VEIN ------- 76
2.2.9 THE TIME AND PLACE TO ASK QUESTIONS ----- 84
2.2.10. SIMILARITIES BETWEEN THE PROPHET DHUL-QARNAYN AND THE PROPHET SOLOMON ------- 85
2.2.11 WRITING AND FATE -------------------------------------- 88
2.2.12 DEPENDENCE-SERVICE ----------------------------- 91
2.2.13 THEY DO NOT BELIEVE ---------------------------------- 93
2.2.14 AND THEY DENIED AND FOLLOWED THEIR INCLINATIONS. BUT FOR EVERY MATTER IS A [TIME OF] SETTLEMENT. (Al-Qamar 54:3) --------------------------- 97

2.3 ALLAH'S BOOKS AND THE HUMAN BOOK --------------- 98
2.3.1 CHANGE -- 98
2.3.2 THE SYMPTOMS OF TAKING THE UNIVERSE BACK TO ITS PREVIOUS FORM ------------------------------- 100
2.3.3 THAT WHICH IS DECREASED FROM THE SERVANT -- 102
2.3.4 WATER -- 105
2.3.5 SERVICE --- 106
2.3.6 CREATED WITH RIGHTEOUSNESS ------------------ 110
2.3.7 QUANTUM MECHANICS OR INNER (INVISIBLE) KNOWLEDGE -- 118
2.3.8 THE INVISIBLE --- 121
2.3.9 THE PROPHET SOLOMON AND IMMACULATE REPENTANCE --- 125
2.3.10 THE PROPHET SOLOMON AND THE SABEANS -- 126
2.3.11 THE PROPHET SOLOMON AND BILQIS, THE QUEEN OF SHEBA -- 127
2.3.12 HARMONY IN BODY AND SPIRIT -------------------- 131
2.3.13 ALLAH DOES NOT HAVE SHORTCOMINGS --- 134
2.3.14 ALLAH'S BOOKS AND THE HUMAN BOOK ----- 137
2.4 THE SYMMETRIC SYSTEM OR ALLAH'S ASSISTANTS -- 142
2.4.1 THE SPACE-TIME CONTINUUM --------------------- 142
2.4.2 PEOPLE AND REPRESENTATIVES ------------------ 145
2.4.3 THE FIRST JIHAD IN THE HEART OF HUMAN AGAINST SATAN -- 150
2.4.4. THE PROPHET MUHAMMAD AND JIHAD -------- 152
2.4.5 THE HEGIRA --- 153
2.4.6 THE MESSENGERS OF ALLAH AND THE HEGIRA -- 154
2.4.7 THE SYMMETRIC SYSTEM OR THE HELPERS OF ALLAH --- 157
2.4.8 THE NAME, THE MOST GRACIOUS ----------------- 160

2.4.9 THE KAABA --------162
2.4.10 TO BE RETURNED TO ALLAH AND HIS NAMES AND TITLES --------166
2.4.11 FORM AND CIRCUMAMBULATION --------168
2.4.12 IRON AND FORCE --------169
2.4.13 THE IMPORTANCE OF AWARENESS --------172

3 HEADING TOWARDS QUANTUM MECHANICS --------175
3.1 ALLAH'S VIEW --------175
3.1.1 HEADING TOWARDS THE SIDRAT AL-MUNTAHA THROUGH THE FATIHA --------175
3.1.2 EVERYTHING, DESCENDING AND ASCENDING ↑ ↓ --------178
3.1.3 THE GLANCE OF ALLAH IS ONE WORD --------182
3.1.4 THE GLANCE OF ALLAH --------183
3.1.5 ALLAH WILL COMPLETE HIS DIVINE LIGHT --------185
3.1.6 FLOWERS OF DIVINE LIGHT IN THE HEART --------188
3.1.7 THE PROPHET JOSEPH AND THE DIVINE LIGHT OF ALLAH --------191
3.1.8 CHILDREN, SCIENCE, KNOWLEDGE AND THE SURAH ASH-SHARH --------195
3.1.9 MOTION --------198
3.1.10 ALLAH HAS GIVEN YOU THE NAME 'MUSLIMS' --------202
3.1.11 HUMANS ARE A PART OF EVERYTHING --------206
3.1.12 ROLAND BARTHES: PUNCTUM AND DETAILS --------208
3.1.13 TRUTH --------212
3.1.14 THE MATHEMATICAL RELATION BETWEEN WORDS --------213
3.2 WOULD THERE HAVE BEEN *BE!* WITHOUT ANY VERSES? --------218

3.2.1 HUMANS AND THE EARTH --------------------------- 218
3.2.2 CREATION AND MERCY --------------------------- 219
3.2.3 MATTER AND SHADOWS --------------------------- 223
3.2.4 THE SURAH AL-MAA'OON AND THE FATIHA --- 228
3.2.5 WHY OBEDIENCE AND FAITH? ---------------------- 229
3.2.6 THE CREATION OF LIFE WITHIN AN OBJECT - 231
3.2.7 THE SURAH AL-ALAQ AND KNOWLEDGE -------- 232
3.2.8 THERE IS ALWAYS SOMEONE WHO KNOWS BETTER --- 237
3.2.9 VERSES, *BE!* AND ATOMS --------------------------- 239
3.2.10 THE SCHOLARS, MUHYIDDIN IBN-I ARABI AND MEISTER ECKHART, AND *BE!* -------------------------- 241
3.2.11 IS *BE!* POSSIBLE WITHOUT VERSES? ----------- 243
3.3 THE MIRACLE OF PATIENCE THAT DECELERATES ELECTRONS --- 247
3.3.1 TABULA RASA: PAPER, PENS, AND WORDS ---- 247
3.3.2 SOLAR SYSTEMS AND REVELATION --------------- 249
3.3.3 GABRIEL, THE ANGEL OF REVELATION, AND THE STAR SIRIUS -- 254
3.3.4 IS THE DARKNESS MATHEMATICAL? ------------- 257
3.3.5 THE SEVEN SKIES ------------------------------------- 258
3.3.6 OBJECTS AND THEIR SHADOWS -------------------- 261
3.3.7 THE PROSTRATION OF OBJECTS ACCORDING TO THEIR CREATION -------------------------------------- 261
3.3.8 THE LANGUAGE OF HEAVEN ----------------------- 264
3.3.9 PRAYERS AND BALANCE --------------------------- 267
3.3.10 RESPONDING TO PRAYER -------------------------- 268
3.3.11 PATIENCE AND TIME -------------------------------- 272
3.3.12 THE MIRACLE OF PATIENCE WHICH DECELERATES THE SPEED OF ELECTRONS ---------- 274

4 THE PARADIGMS OF QUANTUM MECHANICS ----------- 275
4.1 M-THEORY AND FATE ---------------------------------- 275
4.1.1 SCHRÖDINGER'S CAT -------------------------------- 275

4.1.2 AN INTERPRETATION OF EWG/MANY-WORLDS ------280
4.1.3 INTERNAL-EXTERNAL MULTI-WORLD THEORY ------283
4.1.4 M-THEORY AND FATE ------284
4.1.5 M-THEORY AND (SUPER) STRING ------286
4.1.6 MALDACENA AND ADS/CFT APPLICATION ------288
4.1.7 THE BREATH OF ALLAH-1 MATRIX ------289
4.1.8 ALLAH'S BREATH-2 (SUPER) STRING ------291
4.1.9 THE GOOD AND THE BAD ------295
4.1.10 THOSE WHO SOW DISCORD ------300
4.1.11 GOODNESS-EVIL/BIVALENCY ------303
4.1.12 THE NAMES THAT REPRESENT THE OBJECTS BEFORE CREATION ------307
4.2 EINSTEIN'S *THEORY OF GENERAL RELATIVITY* AND FLORENSKY'S *REVERSE PERSPECTIVE* ------309
4.2.1 DARK ENERGY ------309
4.2.2 HUMAN PERSPECTIVE AND THE FIRST HUMAN ------310
4.2.3 THE COLOR AND TONES OF REVERSE PERSPECTIVE ------315
4.2.4 REVERSE PERSPECTIVE AND MULTIPLE WORLDS ------317
4.2.5 EINSTEIN'S THEORY OF GENERAL RELATIVITY AND FLORENSKY'S REVERSE PERSPECTIVE ------318
4.2.6 THE ASTRONAUT, REVERSE PERSPECTIVE AND RELATIVITY ------321
4.2.7 BELL AND NON-LOCAL INTERACTION ------323
4.2.8 NON-LOCAL QUANTUM ------327
4.2.9 CONSCIOUSNESS AND CROSSOVER ------329
4.2.10 QUANTUM AND RADIATION ------332
4.2.11 ZEILINGER AND TELEPORTATION ------334

4.3 ALLAH CAN ABOLISH ANY VERSE HE DESIRES ---- 337
   4.3.1 IS IT IRREVERSIBLE OR AN ECOLOGICAL FOOTPRINT? ---- 337
   4.3.2 IS IT RECYCLING OR BLACK HOLE? ---- 340
   4.3.3 ALLAH CAN ABOLISH ANY VERSE HE DESIRES ---- 342
   4.3.4 GRAVITATION AND QUANTUM ---- 345
   4.3.5 THE EVENT HORIZON ---- 347
   4.3.6 BLACK HOLE AND CREATING ---- 348
   4.3.7 QUANTUM, GRAVITY AND IRON IN SPACE-TIME ---- 349
   4.3.8 THE WAY THE ATOM IS DIRECTED TO MECCA ---- 352
   4.3.9 EMOTIONS, COLORS AND SPACE-TIME ---- 355
   4.3.10 THE CHANGING OF A BLESSING AND QUANTUM CROSSOVER ---- 356
   4.3.11 STRENGTHENING THE BELIEVERS WITH THE ANGELS ---- 360
   4.3.12 RUMI-SHAMS: THE WINE OR THE ROSE? ---- 362
   4.3.13 THE FUNDAMENTALS OF LIGHT, WAVE AND LASER ---- 367
   4.3.14 EINSTEIN: ALLAH DOES NOT PLAY DICE ---- 369
4.4 THE COMMAND OF ALLAH ---- 373
   4.4.1 CLASSICAL PHYSICS AND QUANTUM PHYSICS ---- 373
   4.4.2 IS A THEORY SUBSTANCE ACCORDING TO THE UNDERSTANDING OF QUANTUM MECHANICS? ---- 376
   4.4.3 WAVE-SUBSTANCE AND SUBSTANCE-WAVE ---- 377
   4.4.4 THE PHYSICIST, LOUIS DE BROGLIE ---- 381
   4.4.5 THE TRANSFORMATION OF LIGHT INTO DARKNESS IN THE CONSCIOUSNESS ---- 384
   4.4.6 THE UNITY OF THE WAVE AND THE SUBSTANCE ---- 386
   4.4.7 THE TRANSFORMATION OF SUBSTANCE ---- 389

4.4.8 RUTHERFORD AND THE HALF-LIFE ---------------392
4.4.9 LUDWIG WITTGENSTEIN AND THE CHESS OF DUALISM --394
4.4.10 DIRECTING ELECTRONS AND THE KAABA ---396
4.4.11 THE UNIQUE PRINCIPLES OF THE WERNER HEISENBERG AND THE EWG MULTIPLE WORLD INTERPRETATION---399
4.4.12 THE COMMAND OF ALLAH--------------------------402

5 A HEART FOR SYMMETRY--409
5.1 HANDS HOLDING A PEN AND THE HEART ------------409
5.1.1 QUANTUM AND THE HEART--------------------------416
5.1.2 THREE HEARTS AND A MIND-------------------------418
5.1.3 QUANTUM PHYSICS AND THE SOVEREIGNTY OF THE HEART FOR THE HEART----------------------------------419
5.1.4 QUANTUM, THE HEART AND THE THEORETICAL PHYSICIST WOLFGANG PAULI----------------------------------423
5.1.5 THE OPERA AND THE LOBBYING SYSTEM -----425
5.1.6 GOING BEYOND THE ORDINARY BY TAKING ON RESPONSIBILITY --427
5.1.7 THE DOMINO EFFECT AND JUSTICE --------------431

THE QUR'AN AND QUANTUM PHYSICS

HEART & SCIENCE

THE LEGENDARY PHOENIX

They say a phoenix lived in the land of fairy tales. There had been a lot of people who had made many efforts to reach the phoenix. Yet, the Phoenix did not live in fairy tales, nor would she travel in such distant lands. For her, it was all about one matter, which was to see, for seeing would differentiate everything. This is because nobility is in the seeing and seeing makes the one next to you a phoenix.

There were many beauties in the phoenix... The beauties found within it were to never come to an end. She always waited to be discovered. This is because with each sunrise, sunset and each moment of the day, everything that made her a phoenix would be revealed in her. She didn't know what it was to hide her beauty as if a secret, but rather she would share everything with anyone who wanted to see it.

At the dawn of each new day, the phoenix wanted to reveal its various colors among its feathers that had never been seen before. However, people were never able to do this. They weren't able to look inside of themselves; they were too scared. They never had the courage to ask themselves why they were scared. Rather, people found it easier and more attractive to go on long journeys searching for this beauty in distant lands under every single stone. And yet, the places where the Phoenix stayed were always close to the people and their lives. However, the people who devoted themselves to the search for this artificial love and knowledge actually buried themselves in darkness, not the phoenix. Whereas the moment that men stops being afraid of losing, it is that moment that they will come to the knowledge that there was nothing *to lose. Whenever they distance themselves from taking two values as criteria in a three-dimensional world, it is then that they would avoid judging someone according to those values, and in its stead they would continue their search not in distant lands but with a childish temperament. This was the only way one could encounter the phoenix and it was only at that moment that he might turn into her, and thus, the phoenix would no longer be a myth, because it never deserved to be so.*

1 INTRODUCTION

1.1 LIVING UP TO THE FATIHA

1.1.1 THE FORMULA OF THE UNIVERSE

Since humans have come into existence, the excitement towards *finding the formula of the universe* has not lost and will not lose its appeal. However, people who think that they are very knowledgeable forget that even if all of humanity unites, they can have variable knowledge of the universe, or rather a small percentage (four percent only) of the knowledge reflected in the universe. For this very reason, man does not have the ability to manage this knowledge to the benefit of the people. This is especially true if their heart is left out of the life system. In the Surah Luqman, the prophet explains this in the best way through the advice that he gives to his son. He gave advice to understand the glory of Allah and recognize his own impotence. It is only such a person that desires Allah and, as a result, begins to lift off the secret veils of this desire. This is because Allah does not leave any desire unanswered and he is graceful. There is a prayer in which He promises to reinforce the way to Him, namely the Surah Fatiha.

1.1.2 LIVING UP TO THE FATIHA

I am alive

the place was different this morning

as was what was happening

if you have learned

to live in the past of the future

you will experience the value,

love, affection
of the word given to Truth
that is the Fatiha

and sometimes also
you talk
to those around you
with void words

The Qur'an's first surah and its very essence Fatiha, of course, exists in all deeds. The more intense Fatiha appears in the deeds, the more intense life is found in the hereafter and in the 'Adn, Al-Firdaws and Al-Na'im Paradises. Deeds come forth through the influence of the object. This means that knowledge is hidden in its very object. Therefore, the secrets of the Fatiha, which experiences all its deeds, show that they exist in all beings. If Allah is the Lord of the good and the bad, He shall accept His servant's knowledge of the Fatiha, as well as the prayer he makes out of the love of this knowledge. The purpose is to show the truth in all prayers accepted, so that in return for the knowledge transferred to the object, the object will speak of its own constituents. The multitude of knowledge is only possible through praise, repentance and faith in Allah. Whether or not one has lived for a hundred years, he lives only as much as the truth he has experienced, that is, his faith. Thus, *those who deem attaining to Allah a lie* assume that they live only one hour a day.

Fatiha is a path of seven levels. The beneficial secrets that Allah bestows upon people are hidden in it. It is the power of the Qur'an. It is the key to the door that opens to *al-Lauh al-Mahfuz,* the exalted book, belonging to my Almighty God. It is the first step, the beginning of the prior. And this beginning is eternal. This eternal is Bismillah. It is the letter B in *Bismillah,* and it is the dot on this letter. This dot is an atom. It is the sufficiency of the atom and its particles to form molecules. It is the equilibrium present in atoms with its variability constructing new equilibriums that require gravity. It is the performance of its necessary duty simultaneously that exists by both renewing itself and by the delivery of the particles of the atom to their own time.

1.1.3 THE SPATIAL SIGNIFICANCE OF FERTILE LANDS

Allah sends prophets to the main centers of cities. He does not destroy places where He has not sent any prophets. When knowledge is conveyed about the existence of Allah, this knowledge requires responsibility. Failure in taking this responsibility means denying the truth. The most basic task of this responsibility is to think about what is heard, seen or happening, and then question these very things. Thinking directs your heart as well as your nafs. Thus, the nafs follows the truth. Otherwise, the nafs and the devil deceive people. They actually end deceiving themselves, and their hearts and ears are deafened as they try to bury themselves in excuses. It's such that they come up with the excuse that they are going to be exiled from their homelands if they follow the right path, and the homeland they will be exiled from is the sacred Mecca. Allah has placed them there and all kinds of provision are found in Mecca. During the history of Mecca, the prayer that should be remembered about sustenance and that the Prophet Abraham and his son, the Prophet Ishmael, made to Allah before building Baitullah was as follows:

And [mention] when Abraham said, "My Lord, make this a secure city and provide its people with fruits - whoever of them believes in Allah and the Last Day." [Allah] said. "And whoever disbelieves - I will grant him enjoyment for a little; then I will force him to the punishment of the Fire, and wretched is the destination." (Al-Baqara 2:126) As seen in this verse, Allah accepts the prayers regarding this land. However, He wants people to worship Him and have faith in the afterlife. Otherwise, the tyrant people who deny the truth like the previous tribes would be destroyed. However, the places where believers have settled are safe, and here people are honored to be heir of the fertile land. The spatial meaning of the fertile lands is to maintain social life in welfare and peace without calamities. However, the required resolve of faith is considered difficult for people. In that, first of all, one must touch the verses that are within himself. In other words, he must himself become each verse of the Qur'an revealing the truth within himself, so that Allah will grant him the ability to conceive, realize, bring together the meanings or features of the verses that He brings out of man's depth, in short, let the sound of the Horn make his senses tremble flowing into his heart.

Allah grants the nature of all things and creates the reasons for their existence as much as man reveals the secret verses within him. Existence then gains its meaning as its foundation comes into existence. In general, knowledge is about creating a physical disharmony by doing things such as reading books and life itself, and it has as deep a meaning as this disharmony. The deeper and more meaningful the knowledge is, the quicker it is dissolved into a singular knowledge. In other words, Allah has only one word or even a single glance and teaches His servants to look with this single glance. Thus, His servant sees the true path and gains the ability to see whatever path is bestowed upon him. His dhikr forms rings that make the invisible visible for other dhikrs. But one thing to keep in mind is that His servants need help. There must be helpers so that Allah is mentioned. It is necessary to remember Allah.

1.1.4 HUMANS AND TIME

Man does not settle with just what he sees; time also gives him from itself. Whatever man leaves to time while he runs and chases after something, time gives back to him the moments that have been given to it in space. Only then can human being move these absolute unchanging moments to other spaces and create space-time coordinates. Otherwise, neither man can survive nor the thing or concept that may occur within that space-time can form.

1.2 CONFUTING SUPERSTITIONS

1.2.1 MEANING IN THE KNOWLEDGE OF THE APOCALYPSE

Indeed, the Hour is coming - I almost conceal it - so that every soul may be recompensed according to that for which it strives. (Ta-Ha 20:15)

All people exhibit good or bad attitudes and behaviors correlating with the movement and mobility of their nafs. They therefore pursue implications from good-bad value formats. Since there is no good in this pursuit, it is clear that the human attitude mentioned here is to accept evil. *Have you seen the one who takes as his Allah his own desire? Then would you be responsible for him? (Al-Furqan 25:43)* Therefore, they pay for what they do.

The desire of Allah to hide and hold back the apocalypse is understood in two ways. The first one is that since Allah does not conceal the apocalypse from them, everyone knows that the apocalypse will break forth because how this is going to happen is written in the Qur'an.

Man has this knowledge because he carries the book within himself. If Allah were to hide the apocalypse from him, then man would always be in pursuit of unruliness without this knowledge. The conclusion here is that the knowledge of the judgment day represents truth. It is for their welfare that Allah does not conceal this truth from humans. In fact, they abandon going astray as Allah gives them time. This indicates that the infinite mercy of Allah is found within the knowledge of the apocalypse. Another meaning is that despite the knowledge of truth people who are on this path increase their unruliness. The extraordinary efforts of the Prophet Moses in the stages of calling Pharaoh to the faith can be presented as an example.

1.2.2 THE PROPHET MOSES

When the Prophet Moses was to be born, the priests told Pharaoh that a boy, who would cause trouble for him, would come into world. Pharaoh killed all the boys born at that time. Through revelation, Allah tells the Prophet Moses' mother to put her baby in a basket and leave him in the river Nile, so that he will not be killed. Then, so that she doesn't worry, Allah announces that her child is a prophet and will be given back to her. Even though the mother's feelings did not mislead her, Allah strengthens her heart so that the situation is not made known, as she may appear troubled. The promise of Allah comes completely true.

Pharaoh's family appointed Moses' mother as a wet nurse, and thus, she got the opportunity to unite with her son. Allah then gave Moses knowledge and wisdom during his youth. While he had that knowledge within himself, Moses killed someone by accident. The next day, he almost made the same mistake again. Moses, who had trouble during the night because of the murder he committed, begged Allah to show him the right path. Then, Allah led him towards the right path delivering him from his fears. For the Prophet Moses, the right way meant Midian. Through begging of Allah, he found this path, which gave the Prophet a peaceful ten years, a wife and a son at peace. At the same time, during that ten-year period, the Prophet Moses regained his strength to stand against Pharaoh and to bear the burden that was waiting for him in the future. Peaceful or difficult times experienced whether for a short or long period of time are the means by which Allah tests one in order to strengthen his faith and/or to see if he is able to overcome his burdens.

1.2.3 THE VIRTUE OF LOVE

As in the qisas of Pharaoh and the Prophet Moses, it is so obvious that Allah wants to forgive people from paying for their sins, as long as the servant takes a lesson from his deeds, because Allah is merciful.

Allah asks Moses what he has in his hand, as he holds his staff. With this staff, the prophet would accomplish his minor tasks and see some of his other needs met. The staff is connected to him and bears the knowledge of his characteristics.

Thus, Allah anoints the staff with two miracles in order to ease the Moses' tasks. The first miracle was that Allah anointed the staff with main characteristics of magic itself. The second miracle was that, upon Allah's orders, the Prophet Moses put his hand under his arm and his hand revealed a flawless and spotless whiteness, that is, a divine light. The presence of these two miracles assigned by Allah within the Prophet Moses was necessary for Allah to show other verses from this place. In other words, when a person becomes aware of the miracles, talents and skills within him and applies them to his life, then Allah reveals to him other verses. This means that they way you look at life, that way it becomes, or rather you will reap what you sow.

According to the degree of difficulty of the task, Allah may also support His representatives by other means beyond his verses. The Prophet Moses was afraid of the great task that he had been assigned against the cruel Pharaoh, who had declared himself as Allah. In order to overcome these fears and his task, the Prophet asked Allah to assign his brother Aaron to help him, mentioning about his characteristics. In that, he suggested that his brother's speech was better than his, and thus, he could talk more eloquently about Allah. In this manner, Allah chose the Moses' brother Aaron as a prophet as well. Thus, by demolishing all the negative attitudes and behaviors of the pharaoh, the two brothers brought about peace in place of everyting caused by such negativeness.

Allah gave the Prophet Moses self-love as a protection against his fears. Characteristics such as peace, confidence and power are the features that one, who has love and affection towards his surroundings, finds actually within him. Where love is, evil turns into good. This is because where love is, Allah assigns His helpers as guardians for the dwellers in need whatever the matter is. That is, Allah sent the Prophet Moses and his brother

Aaron as a support against all the malignancy that the Pharaoh represented and brought about. In addition, being enriched with symbols such as blood or magicians repenting in *"the mid-morning on the day of the festival,"* this anecdote pointed to the omnipotence of Allah.

1.2.4 CONFUTING SUPERSTITIONS

The struggle between the Prophet Moses and Pharaoh began with the tricks of three magicians gathered by Pharaoh. The three magicians threw their staffs to the ground and these staffs magically appeared to be snakes. They couldn't get beyond just an illusion, however. When it was his turn, the Prophet Moses tossed his staff to the ground, and the staff swallowed up the things the magicians had concocted. What manifests itself in the hereafter (worshiping idols) manifested itself on earth. Thus, the truth was revealed and the things the magicians had done were vanquished.

The three magicians thereupon fell down in prostration and said, *We have believed in the Lord of the worlds, the Lord of Moses and Aaron (Al-A'raf 7:121, 122).* The magicians immediately realized the truth of this single event that resulted in the confutation of their deception, that is, superstition. Other people at this meeting couldn't see what these three magicians saw. Other than the two prophets, the magicians had seen the truth and how it had destroyed their deceit clearer than anyone else there. In the face of this pure truth, they didn't give up their faith in spite of Pharaoh's threats and wrath. Along with this, Allah sent many calamities upon Pharaoh one after another, but he still insisted on not believing. As he was about to drown alone in the sea, he declared his faith in Allah at his last gasp, yet it was already too late.

The differing perceptions between Pharaoh and the three magicians explain how the superstition has been confuted. On one hand, there is a tyrant, who declared himself a God and who ordered people to worship him. On the other hand, there are three ordinary people who have taken up the profession as magicians in a period when magic was widespread. People learned magic from two angels named Harut and Marut during the reign of the Prophet Solomon. This was a test for people, but they abused magic. Therefore, Allah forbade magic and cursed those who practiced it.

1.2.5 GOODNESS AND EVIL

Except for those who repent, believe and do righteous work. For them Allah will replace their evil deeds with good. And ever is Allah Forgiving and Merciful. (Al-Furqan 25:70) Why are there verses related to magic, gambling and wine in the Holy Qur'an? Why does Allah inform people about the existence of these things and about the wisdom that may be related to them? There is no distinction between good and evil in the unity dimension of Allah. But the Qur'an serving as a mirror for each creature within this universe tells people what good and evil are and what people should pay attention to in order to walk on the right path. The situation, which should not be forgotten when speaking of objects, words and concepts, is the fact that Allah has created man asymmetrically through the words of this dualism. In other words, humans do not only have the potential to conceive good and evil. It is also necessary for the concepts of good and evil to be created in his mind so that the potential to perceive the good and evil can be imagined and put into practice. Moreover, nothing is completely good or bad. It is not possible to grasp this because human being is adamite. Therefore, one cannot afford to conceive this, for he requires the power to create. Creating a speck does not only mean bringing it into existence. It also means incorporating it into all creatures, spaces, and knowledge outside of creation. The supreme Lord, thus aware of the speck, is the sole owner of the ability to create. *"O people, an example is presented, so listen to it. Indeed, those you invoke besides Allah will never create [as much as] a fly, even if they gathered together for that purpose. And if the fly should steal away from them a [tiny] thing, they could not recover it from him. Weak are the pursuer and pursued." (Al-Hajj 22:73)* There is one more thing that points to the power and knowledge of Allah and that is the structure and functionality of quantum mechanics, which are based on possibilities. Within the context of good and evil, everything can be turned into as much good as it is evil. Otherwise, it wouldn't be necessary to repent. *Say, "What would my Lord care for you if not for your supplication?" For you [disbelievers] have denied, so your denial is going to be adherent. (Al-Furqan 25:77)* One should also take into account margin of error. *Fighting has been enjoined upon you while it is hateful to you.*

But perhaps you hate a thing and it is good for you; and perhaps you love a thing and it is bad for you. And Allah Knows, while you know not. (Al-Baqara 2:216)

1.2.6 THE FEAR OF ALLAH

Such is the admonition given to him who believes in Allah and the Last Day. And for those who fear Allah, He (ever) prepares a way out... And whoever relies upon Allah - then He is sufficient for him. (At-Talaq 65:2, 3)

Whether a person does trust or not, or fear Allah or not, he will fulfill His commandments eventually. In order for this to happen, Allah has put a measure into all things. Within this context, it is necessary to fear and rely on Allah rather than standing against Him. Because those people who know that Allah will fulfill whatever they want, nothing will be difficult. Even if it is difficult, this pessimistic viewpoint will be replaced by an optimistic approach when these things are trusted. The servant, who possesses this way of thinking, connects the two worlds through the doors opening up to him, and as a result, his degree increases in both. For example, if a family does not have any children, and they still appreciate Allah's commandments avoiding to rebel against Allah for this, they gain virtue. Allah grants this faith in exchange for patience. Maybe He will never give that family children, but in exchange for their patience, He will open the doors of the world and the Hereafter unimaginably. He makes their life meaningful in terms of truth.

Again, Allah lightens the burden of those who fear Him and covers their wickedness. Even, he turns their evil into good, and their reward increases just like a spike grain. To fear Allah means to follow His commandments. The very foundation of these is to be just.

One who fears Allah is always intellectually active, because he is always in a position of self-reasoning in consideration of Allah' approval. Trusting Allah means to comply with all things experienced, to remain in tranquility, to be patient, to be completely surrendered, and to fulfill Allah's commandments. In this situation, one is passive.

2 SHAPING TIME ACCORDING TO SPACE AND KNOWLEDGE

2.1 OVER EVERY POSSESSOR OF KNOWLEDGE IS ONE MORE KNOWING

2.1.1 HERMENEUTICS (INTERPRETATION)

The word hermeneutic comes from the Greek God Hermes, and in the 19th century, it began to be used for the theory of interpretation and understanding. It is a scientific method that defines things and human beings in the form of words and that aims to reach their meaning, i.e., the truth. Most of the narratives in the Qur'an repeat themselves but always from different angles and directions and in different harmony (the Holy Qur'an's mathematics and Aleph) bringing them to diverse wholeness.

2.1.2 DETERMINISM

And Our Command is but a single (word), like the twinkling of an eye. (Al-Qamar 54:50)

According to determinism, predetermined laws guide the universe and its system. Therefore, it is foundational in this view that events take place outside the will of the human. The word of wholeness or singularity that corresponds to determinism in the Qur'an is *Kûn* (be). In this context, it is clear that the singular will exists. If there is a singular will, then everything is constituted of atoms according to determinism's application field within the laws of physics. Allah has given man the duty of revealing what exists. The creation of man is perfect and his asymmetry makes him even more perfect. In this regard, Allah has granted him a purpose and has given him reasons to achieve this goal. In other words, humans are asymmetrical; that is why they are not perfect. But if there is no asymmetry, the necessary energy, which is also necessary to live, cannot be achieved. What both ensures these and is they themselves is nafs. Everything that happens within man and the universe can only rotate around its axis and revolve in other orbits through this asymmetry. Thus, the survival and way of life of all people corresponds to recurrence. This has to be so, in this way, it can bring itself into existence by drawing circles... In other words, the existence of both depends on each other. The line of truth may only appear in a good-bad reaction. And the more truth is discovered and experienced, the more moral beauty desires to be repeated. With the dimensions of moral beauty, truth becomes more visible.

2.1.3 REVELATION AND THE CALL

Move not your tongue with it, [O Muhammad], to hasten with recitation of the Qur'an. Indeed, upon Us is its collection [in your heart] and [to make possible] its recitation. So when We have recited it [through Gabriel], then follow its recitation. Then upon Us is its clarification [to you]. (Al-Qiyama 75:16-19)

The revelation of Allah appears to all things and beings in the world in different ways and forms. To understand them, Allah has given people sense organs in order, for example, to see, to hear, and to feel. And, He has created them to the extent that they all affect each other and become integrated, so that they head towards the depth of the verses and towards a single word according to the acquired consciousness of the hereafter. Hence, the layers of the senses have been created according to the seven-dimensional universal leaps and transitions from one realm to another. This begins with the separation of the body while living in it. During this decomposition, there must be the necessary amount of chemicals that the body needs. For the one who has set off to find Allah, only in the case when the knowledge of how much and what is given to him within this journey and of what these particles bring flowing through the body can the balance be established in all conditions. Rumi's words summarize this adventure of separation: "I was raw, I got cooked and I got burnt."

In this world, the rule of truth can only be realized in such a way. This can be illustrated by addressing the relation between the mind and the heart within the body, which is much stronger than gravity. This more powerful thing is the Horn. It is important to note that Allah is closer to human being than he is to his jugular vein. Allah helps man in every situation because He is merciful. And then, man turns to Allah and recognizes his Creator. But it is not that easy for man to reach Allah. This is because only the truth has the right to do so. For this, the more truth dwells in the blood of man, the closer one gets to Allah. Then, Allah draws that servant closer to Him to the same extent. All the prophets exemplify the most beautiful positions that human beings have obtained. Among them, Prophet Muhammad met Allah at Sidrat al-Muntaha. He completed his journey in this world before he departed it. He went to places where Allah had sanctified, such as al-Masjid al-'Aqsa and al-Masjid al-Haram. Then, from there he went towards Allah. In order for him not to lose his way, he must have already had the knowledge of this direction within him. That is, the seven-layered veils—ambivalence—directed towards the Sidrat al-Muntaha should be lifted so that he could reach his destination. What kind of knowledge of the truth was it that while still alive?

2.1.4 SHUAIB: THE PROPHET OF JUSTICE IN THE WORKPLACE

Everything has a foundational meaning as much as it has other meanings. Foundational meanings are the starting points, the ropes that lead us to Allah's point of view. The measure is shaped according to the movement and mobility of the nafs, the dispersed particles of gravitational force, or the action and reaction of the functional variability. Their foundational meanings are then scattered throughout other various meanings. Deviations due to the matter's asymmetric structure attract wrong behaviors and wrong decisions and they end up becoming the universe's and the person's own geometry as if experienced traces falling into a magnetic field. If the measure lacks fairness, then the balance that Allah provides and requires of people is disrupted. Positive or negative, the balance is always there. However, the important thing is the line of measurement put forth by Allah. That measure covers all things, from internal affairs to outwardly attitudes and behaviors. Yes, a human is mortal, but a mortal with a heart. Thus, Allah's measurements have been entrusted to the heart. As greetings of immense serenity within the heart come together, they have become as much in love with each other as the love that the heart has for them and to the extent of their knowledge of Allah. They are divinely nourished by every truth that is evident, as is the heart. And here writes heart, which is the only heir to be desperate for the love of Allah. There is no lie in the heart's pen but only the words of Allah. Allah's name, *Just*, flows through the center of His words for the measurement to function greatly.

However, to the extent that a disruption occurs in the ascent to and descent from the heart, remembering Allah in our conscious breaths, or unconscious breaths that come out of our consciousness, and being aware of His justice regress at the same rate. To illustrate this from the viewpoint of social life, what is written about today's history is quite obvious. There are socio-psychological crises resulting from the construction of buildings and the lobbying system on the one hand, and on the other hand, due to wars, poverty, and consumerism that innocent people have been exposed to. They depict all the space-time circles and middle diameters that must be in space-time as if they were distant from the lines of the heart.

2.1.5 THE PROPHET SALEH AND THE SHE-CAMEL REPRESENTING CREATION

Allah sent down a she-camel to the non-believing people around the Prophet, because the people of Saleh did not comprehend the knowledge necessary for faith in Allah. This information could also have been met by recognizing the features of the camel and drawing a lesson from these characteristics. The characteristics of the camel were countless, but the most obvious ones for faith were that the camel can live without eating or drinking for a long time and she is very resistant to the cold. Whatever task she is given, she recognizes it and obeys, and lastly, she is fertile. In other words, only such people who are so faithful to their fate and act according to the truths of their disciples are scholars who can give birth to those with whom shares the same features.

2.1.6 ARE THEY MORE THAN NECESSARY, OR ARE THEY DEFICIENT?

The person who lives in interaction with his endless desires not only adds his voluntary or involuntary differences to his inner world but also to the universe. They are the features of that person becoming visible by touching those differences in himself with the differences he has brought and handed over to the universe.

Therefore, based on these differences, the activity of these objects as well as himself increases in the same proportion.

2.1.7 HORN-TRUTH-THE DAY OF JUDGMENT

What is the horn? The Surah Al-Inshiqaq states that the sky listens to its Lord at the moment it is split. With the splitting of the sky and the abolishment of the evil from the earth, the truth completely reveals itself. In other words, the breaking forth of the apocalypse means turning the asymmetrical creature into a symmetrical state or passing through this realm into a completely different realm. Neither humans nor the universe can afford to carry this truth, because the universe is composed of atoms and the human is also asymmetric. Therefore, when the apocalypse breaks forth, the universe will be in tatters and everything that is alive will die. After this death, when Raphael (PBUH) blows the horn a second time, the resurrection will begin, and the more the human possesses the truth within himself, that is, the more he has obeyed Allah and the Prophet, he recognizes the truth. As he gets more familiar, he will feel safe and the Lord will help his servant so that he does not feel alienated. Allah gives His servant his book from his right side. In the Surah Ya-Sin, Allah depicts the reasons that people feel alienated against the truth, and accordingly, in which manners they will appear using the terms *low and worthless*. He who does not prostrate himself in the world will not prostrate himself in the Hereafter either. This is because in the hereafter, they cannot find fulfillment for their emotions, behaviors and prayers, of which they haven't fulfilled in the world. What man has fulfilled in the world before he dies goes on before him into the hereafter. In fact, this means that one constructs the blocks of his heaven or hell for the hereafter. Surely, the Surah Al-Maa'oon shows the way to heaven. For this, the servant is invited to prostrate himself before the hereafter. The remaining sins of he who is able to prostrate himself are covered up. The one who feels a sense of strangeness towards prostration will bring upon him his own hell. Prostration is not only the forehead touching the prayer rug, but rather is the entirety of all attitudes and behaviors done for and pleasing to Allah.

2.1.8 WITNESSING

And [mention] when your Lord took from the children of Adam - from their loins - their descendants and made them testify of themselves, [saying to them], "Am I not your Lord?" They said, "Yes, we have testified." [This] - Lest you should say on the day of Resurrection, "Indeed, we were of this unaware." (Al-A'raf 7:172)

The above verse in Surah Al-A'raf of the Qur'an clarifies the question of determinism pertaining to physics and philosophy in particular. There are two important things to ask here. First, why does Allah make it necessary that He Himself will be a witness against man? Second, what does being a witness here mean and how can it be achieved? When we begin with the second question, being accessible means being able to go from one place to another. Reaching these places can only be possible through the means and will of the nafs.

The nafs takes the mobility and action needed for energy from the existence of dualism (good-evil). The notion of good-evil here is always relative as is Einstein's theory of relativity. This is because the need for variability arises due to the relativity of good and evil, and therefore, variability itself arises. This is one of the first conditions for mobility and movement to continue; in fact, it is the first necessity. When we go back to the first question (why does Allah make it necessary that He Himself will be a witness against man?), we see that this is a necessity for human beings. It is the reflection and reverberation of truth, beauty, light. In other words, it is the presence of Allah's names and attributes in man, in human essence. Furthermore, creatures come into existence because the presence of Allah is in the living creatures that make up the rotating universe. From the perspective of witnessing, this is a situation similar to what is found in the verses of Surah Al-Imran. Allah gives books and wisdom to all prophets. In return for knowing Allah and carrying the character of prophecy, when the prophet Muhammad came, Allah asked them,

Have you taken up the responsibility of my testament? Allah received the word, *We have taken it up* from all the prophets. Allah declared his purpose to the prophets before the universe was even created, that is, since Qalu Bala. They promised that they would perform basic prophecies in order to aid the Prophet, who is the last and the complement of all prophets, and the religion of Islam, which is the complement of all religions. However, there was a sense among the other prophets that the last prophet had not yet come during this meeting in Qalu Bala.

Another witness is the one that is comprised of people in terms of Mussulmanism and Islam. In the witnessing that pertains to Islam, the prophet Muhammad is the witness of all Muslims and of Islam itself. All Muslims - including all the prophets - have been identified as witnesses of humanity and human beings.

2.1.9 THE NOBLE ALLAH AND THE UNKNOWN

Man has been created with superior virtues. Despite this, he knows only four percent of the universe's systemic existence. In the twenty-two percent of it, there is very little known about the universe. Specifically, the act of uniting the unknown with things we know little about can be called entropy in physical terms. The remaining seventy-four percent is completely obscure and unknown for the human being. This fact, on the one hand, points to the various diversities flowing within human beings themselves and that this flow continues internally. On the other hand, man's ability to comprehend is completely limited to these known and little-known parts. These little-known actors that play the role of mediators, in conjunction with the parts of the unknown, mean that there is something more than just man's ability to comprehend with every moment.

When all current movements of this scientific discovery in the social arena are taken into account, there is a lack of expression and the components of this lack of expression depict true expression. The deficiency in the wholeness can be found by understanding of what is going on and putting the missing parts together.

2.1.10 OVER EVERY POSSESSOR OF KNOWLEDGE IS ONE MORE KNOWING

But over every possessor of knowledge is one [more] knowing. (Yusuf 12:76) If it were not so, then the stages related to time and their return to space would never have happened. It does not seem possible to envision a space where there are no stages and thus can be transformed back into time. However, something can be underlined here, which is most fundamentally, there must have been a purpose for Adam and Eve to be sent from heaven to earth. In short, without dualism, humans would not have been able to find themselves nor would they have taken their share of the good and the evil in this life. It is during the very existence of these stages that Allah declares himself. But there is one issue arising from the propagation of existence, in that clear knowledge has become complicated. Allah already mentions in the great book that previous tribes were always stronger, more advanced and more capable. Who knows, maybe the highest age of civilization was the period of Noah who lived for 950 years.

Cities and pyramids that emerge during archeological excavations show that an extraordinary knowledge of mathematics and geometry were used in their construction. What kind of knowledge was this? There is clearly a superior reasoning, but without knowledge of Allah and faith in Him. The fates of all of these groups have resulted in a very bad ending. The person, who is gone forever, carries the knowledge of those who have been subjected to these disasters, because this knowledge has been evident. Therefore, he visits these tribes *every morning and evening.* In the Surah as-Saffat, Allah first greeted all the prophets, then said that people visited in the morning and evening. Therefore, just as the more people do bad things the more they partner with evil; the more they do good deeds the more they get closer to the representatives of that good and receive their shares in glory. And, of course, these tribes will accompany them more *every morning and evening (as-Saffat 37:137, 138).* What emotions do people experience in the morning? Fear? Sadness? In this sense, emotions lead man to different places and times. The important thing is which feelings are leading people and to which period of time and place.

2.1.11 THE EVIL IN UNNECESSARY REQUESTS

But [insolently] they said, "Our Lord, lengthen the distance between our journeys," and wronged themselves, so We made them narrations and dispersed them in total dispersion. Indeed in that are signs for everyone patient and grateful. (Saba' 34:19)

As stated in the aforementioned verses from Surah Saba', Allah provided great opportunities for His servants to live together peacefully. But their insatiable wishes have done nothing but waste themselves and make their lives harder. Yes, the main theme of negativity caused by thoughtless wishes corresponds to the fact that the space-time, which makes up a particular situation, causes them to multiply through various ruptures in space-time. From one perspective, the natural environment becomes increasingly invisible. With the removal of Allah's veils, one will have difficulty depending on space-time, the increase of the existing ambivalence, or the number of fragments. This means that the person becomes artificial to such an extent.

From a more physical perspective, it is absolutely toilsome to reach from one place to another and the culture of sincerity and solidarity between them decreases to the extent that people distance themselves from each other. With this decrease, social-cultural and religion-racial differences increase depending on the region in which they live. The progress in scientific disciplines, the sub-units or sub-branches within themselves, the rise of technology, having different religions, sects and even local traditions in a country depict a starting point that will lead to the diversity of a tiny desire. In short, an increase in diversity means the multiplication of dualism.

Each increase in diversity implies that artificiality will also increase. The proliferation of artificiality means an increase in worshiping and being worshiped. The reality is that it has become more difficult to achieve knowledge of the truth. The reason for this is the increase in the distance between the two-valuation structure resulting from the dominance of the two values and the truth in the measurement of all geometric shapes in the human structure in the macro-cosmos, which is not only associated with place-country as in the previous situation. But whoever is very patient (as in the example of the Prophet Job) and who is very grateful (as in the example of Prophet Solomon) will learn from what is going on around him and will end up understanding better. The weak and foundationless structure, which derives from ambivalence causing people to forget their own will, may present error as truths. Man should avoid making friends with the devil. *The example of those who take allies other than Allah is like that of the spider who takes a home. And indeed, the weakest of homes is the home of the spider, if they only knew. (Al-Ankabut 29:41)*

In other words, the multiplication of dualism serves the devil. When Adam was created, Satan did not obey the command of Allah and did not prostrate before Adam. In return, Allah expelled him from his assembly. Satan promised Allah that he would pervert His worshipers and deceive them. And Allah justifies Satan. However, Allah is the guardian of all things, closer to man than even his jugular vein. His gaze is already present everywhere (as in the matrices in quantum mechanics). Satan, on the other hand, has no influence over people. People's lives should have an essential purpose. Here, Allah gave this opportunity to Satan that he would distinguish the believers in the hereafter from the rest who doubt.

2.2 THE SURAH AL-KAHF AND THE WINDOWS OF KNOWLEDGE

2.2.1 THE SURAH AL-KAHF AND THE WINDOWS OF KNOWLEDGE

Allah reminds us in many verses of the Qur'an that He hasn't created anything on earth without reason, nothing just for fun. Everything has been created in a way to test how His servant will act. But all this will be lost when the apocalypse breaks forth. The Surah Al-Kahf depicts Allah narrating four different events regarding virtuous deeds never being in vain. Although these four events are addressed to different people, they are integrated according to different subjects though they have a point in common, which is knowledge and the knowledge comprehension of faith.

2.2.2 THE OWNERS OF THE GARDEN

Surah Al-Kahf's first story begins with a dialogue between the "owners of the garden." In this conversation between two people, the difference in the people's faith is depicted. The one who lacks the knowledge of faith feels that Allah will love him in all situations, both in this world and in the hereafter, and trusts that he will always be served. And by trusting, he believes that life and everything else will be in accordance with his own will. This is an understanding that contradicts the stages of creation. This is because with the knowledge of the good, evil and the truth, both people and things rotate on their axis and revolve around each other. In this manner, one provides the maintenance of another's mobility. In other words, things and people are responsible for each other. Regardless of whether they want it or not, they do all their rotations for Allah. However, the aim of things and the man with a sound hearth and good sense is to rotate voluntarily for Allah, that is, in the knowledge of Him. Therefore, this trust should not be confused with relying on

Allah. This is because worshiping means to settle for things that are beyond them, to be grateful in every difficult situation, and to hope for Allah's help.

Acting with the knowledge of contemplation and trusting, where one believes that everything and everyone actually revolves around him is not shirk. And, that is why the other person in the above-mentioned story warned his friend in this sense. Thus, instead of being grateful for the eternal blessing that Allah offers, it has been declared that the price of shirk is definitely paid. Moreover, those who do not commit shirk know from where and how the disaster will come, because they understand the situation through the heart. In the end, the Qur'an illuminates the situation with the examples it gives in this story and elsewhere. The pestilence in the area where there is shirk occurs mathematically and chemically. This is a blessing from Allah for those who believe in order to strengthen their faith.

2.2.3 GRACE

As in all things, there is blessing in the totality of grace, which is to reach the knowledge of Allah through the knowledge presented by Allah. Allah has not granted Satan and his descendants this grace. This is because He hasn't made them a witness to the creation of anything; they have been even deprived of the testimony of their own creation. In fact, if they had witnessed their own creation, they might also have witnessed the creation of things. But this is not the case. In Surah Al-Araf, Allah says that He will bring forth people's descendants and make people witness their faith in Him. He says that He has done this in order to prevent people from going through denial, saying that they were not aware of this situation. The conclusion drawn here is that the knowledge of Allah begins with the witnessing and will end in witnessing once again. Witnessing is the grace of Allah's declaration that He has created everything unique, out of nothing, from the realm of the invisible world.

2.2.4 THE GRACE OF FRIDAY

The grace of Friday, which lasts for seven days until the following week, is a completely different gift that Allah has given His servants. With this grace, the heavens race trying to fit themselves into Friday without needing to go on the mirage or search for a way to Allah. Those, or their particles, that cannot fit into this time period on Friday end up drifting into the orbit of other days. On Friday, the angels have the joy of carrying the grace of Allah to earth. In order for the dead to have courage in face of their fear, particles of divine light descend from the hereafter like stars. By the grace of Friday, the gates of delight open up every Friday to the places where the servants have never hoped or even expected. Friday is a body, and as everybody has a heart, so does the miracle of Friday. By this grace, with the withdrawing of time and place, faithful Muslims have the opportunity to prostrate themselves before Allah together with the prophet Muhammad (PBUH) and be successors. *And Peace on the messengers! And Praise to Allah, the Lord and Cherisher of the Worlds. (As-Saffat 37:181, 182)*

2.2.5 THE KEYS TO THE INVISIBLE REALM

The same verses that are found in various surahs of the Qur'an are also quoted in other surahs, or the verse may be quoted again in the same surah. One verse may point to another later verse, or vice versa. The keys to the invisible realm are with Allah. Those keys are partly (i.e., quantum-ly) to locked doors that open up to unrevealed knowledge. These locked doors are found in the heart. If the servant does not persist in disbelief and if he is seeking knowledge, Allah slightly unlocks the secret to him during his search of knowledge. The doors open up as much as one plunges into the comprehension of knowledge.

The keys to the invisible realm lie with Allah and they unlock the secret gates. Obtaining these keys is only possible with dhikr, which may be in the form of reading the Qur'an, praying, remembering Allah by His names and titles, giving material or spiritual alms, giving alms to Allah, or performing the prayers. In short, making mention of Allah everywhere and in every circumstance means to obey His commands gaining his approval.

The secret being evident is a disparate movement emerging from a time-place mobility that leads to movement within the atom from the perspective of physics, that is, the truth. Thinking only means nothing. If thoughts flow in the veins together with the heart, then submissive faith appears on the most righteous path. It is only possible with the heart to hold things in this balance, and Allah has only created the universe for the heart. *Are those equal, those who know and those who do not know? (Az-Zumar 39:9)*

Allah advises His servants who believe in His prophet to avoid opposing to him and to be patient in their difficulties. For this, He promises them unexpected rewards. The responsibility of the believers is to trust in and surrender to Allah and do good works in the world. Thus, they find goodness in return for these good works. However, if the servants are prevented from having a good attitude and behavior, then they may emigrate elsewhere. Remaining and being patient with the events happening around them open the door to uncountable rewards, while leaving is also a solution and permissible. A different reward is given to the former. Although innumerable blessings can be counted and expressed through reason and the heart, one still cannot even guess with his incapable conscious what a blessing it is to be healthy, have children, be economically independent, and be in paradise in the hereafter.

The importance of patience here is specifically emphasized because patience opens up many doors to approach Allah removing people from the monopoly of time and place. With the endurance that one gains by paying the costs, he attains the endless forms of faith and knowledge by the will of Allah. In addition to being a way to draw lesson from the universe, the Qur'an, its being a healing guide and the influence that it has left on man also show that it has given man everything he needs through Allah's words.

2.2.6 VERSES AND SUSTENANCE

On the earth are Signs for those of assured Faith, as also in your own selves: Will ye not then see?

And in heaven is your Sustenance, as (also) that which ye are promised. Then, by the Lord of heaven and earth, this is the very Truth, as much as the fact that ye can speak intelligently to each other. (Adh-Dhariyat 51:20-23)

There are such believers who avoid rebelling against Allah and put their trust in Him. They sleep very little at night, beg for forgiveness and give alms. These are the ones who have an uncompromising faith. After the verses above, Allah mentions the Prophet Abraham and heralds the Prophet Isaac. In this good news, all the bodies in the sky express something that Allah's word has determined for those things and they bear the very essence of what they are representing. For example, the Pleiades, which consists of several stars in the sky, helps us to find the North Pole. The star Sirius is the brightest one in the sky and carries the dimensions of the word *hope* (the string that expresses the degree of freedom in quantum physics).

The synthesis of all the verses on the earth, in the soul, and in the heavens is sustenance. A *verse* means a different harmony of thousands of revelations. With the transformation of revelation into knowledge, harmony moves away from its very essence combining with the artificial. In revelation that transforms into knowledge, nafs inhales knowledge from the revelation with the addition of human desires to the nafs. That is why Allah says that people in the past have always had better knowledge.

There is also truth and revelation in the nafs. In other words, the truth emerges from two values (good and evil) and then the light of revelation shines. Allah stipulates believing to be able to see and comprehend the verses on earth. In fact, believing here is to turn the verses into the knowledge, namely into words, besides preserving their harmony. If believing in Allah did not call for such rebellion in individuals and did not create a reaction in them (which is already considered ignorance), except for Allah's own will (natural disasters), one would see that Allah has made the earth to service him and that He has made it obedient to him. What is in the sky? On which level of the sky or skies are the angels? Does each of them pass through the stars? Do the stars inform that the earth will be shaken by the commandment of Allah?

2.2.7 THE YOUTH OF ESHAB-I KEHF

The Eshab-i Kehf consists of several people who believed in Allah. These people had to leave the place where they had settled because of the infidels of that time. Yet, in return for their faith, Allah took care of their problems and honored them with the knowledge of His own existence. Allah put the people of Eshab-i Kehf into a deep sleep of three hundred and nine years. Thus, he put a veil over their ears and told the Prophet

Muhammad, *If you had seen them, you would have run away without looking back.* Are prophets or ordinary people afraid of skeletons? What kind of a physical change did this sleep require then? What can be taken from the bodies of the young sleepers at Eshab-i Kehf and what is left in their bodies in order not to depart this life?

Another mystery is in regards to the number of Eshab-i Kehf. Allah warned people not to enter into discussions about their number. On the other hand, He pointed out that there are only few scholars who know their number. The sun heated the cave up in a way that it did not disturb the youth. The youth of Eshab-i Kehf knew that the promise of Allah was the truth. Keeping this in mind, they declared this promise. In conclusion, there is knowledge where there is faith. Therefore, the shelter of the Eshab-i Kehf cave was the very path and hiding place where Allah had directed them. This is similar to the case of the Prophet

Moses when he went to Midian, afraid of being killed and lived in that city for ten years with the Prophet Shuaib, or when the Prophet Joseph remained in the dungeon for many years to protect his soul; or the way the Prophet Abraham separated from the people who tried to burn him... Allah bestowed upon them His grace giving them family, children and a home. In the continuation of the story of Eshab-i Kehf, these young people woke up aware of what had happened to them. As they asked each other about how many years they had been sleeping, it was obvious that they knew Allah had let them sleep to protect them.

2.2.8. THE STORY OF THE PROPHET MOSES AND THE PROPHET KHIDR AND THAT ALLAH IS CLOSER TO HIS SERVANTS THAN THEIR JUGULAR VEIN

In the Qur'an, two fish are mentioned. One is mentioned with respect to the events after the

Prophet Jonah was furious with his people and abandoned them. When the ship that the Prophet Jonah was on was about to sink, lots were drawn as to who would be thrown off the ship. The lots fell to the Prophet Jonah. The first fish referenced was the one that swallowed him as a result of being thrown into the water due to *losing*. The fish was big enough to swallow a human being. In a sense, this fish represents the greatness of blessings and grace that Allah gave to the Prophet Jonah. The Prophet Jonah abandoning the place where he was is the very denial of these blessings. However, the denied blessing *swallows* people this time imprisoning them in itself. The second fish was a fish symbol that took a role in the reunion of the Prophet Moses and the Prophet Khidr. The second fish was dead and symbolized sustenance as in the story of the Prophet Jonah. The Prophet Khidr, who enabled this fish to survive in the sea, took hold of that environment through his identity as he was around it. By acting on this magnetic task, he transformed the quantum particles of that area according to his own identity. Both fish symbols could basically represent sustenance. The issue here is that a dead thing comes back to life again and transforms into its prior living state. Thus, what human beings have are the blessings given by Allah, which keep them alive.

That the blessings remain alive is to know that everything comes from Allah and will return to Him again. This is patience, contemplation and the servant's closeness to Allah. Any rebirth or resurrection throughout the entire universe can only be possible through patience, contemplation and trust. The explanation of knowing that everything comes from Allah is given through the story of Moses' journey with an honorable wise man. The main plot of the story is that the prophet was aware of the presence of a more knowledgeable man than himself and sets off with his young assistant to find him. Their journey is said to be quite exhausting. Various ideas have been put forward up until today as to whether the man the Prophet Moses was seeking was a prophet. The Prophet Khidr is actually a representative of time because of his ability to transit between various times in the space-time continuum. Although his name is not mentioned in the Surah Al-Kahf, the person spoken of is supposedly the Prophet Khidr.

Allah's gift to the Prophet Khidr equipped him with innate science. What kind of knowledge was it that the Prophet Khidr could be in many places at the same time? Is that the time that flows as per the view of the people or rather is the space where there are the items and the house in it? In other words, the fact that space could be lost means that either they could be connected to something else or that space could be just a hallucination. It will not be wrong to say both options are correct because man only knows two to three percent of his own mind. Moreover, it should be noted that the parts about which we know well, very little or nothing are separate. In addition to contributing to the functionality of other mechanisms or organisms as known-unknown, they also serve their own integrity for the functionality of integrity in mind.

Without further going down a rabbit trail, we can sum it all up as follows: Spaces definitely carry the knowledge of transforming into something in a way to respond to Allah's command (His command of *Be!*). The presented situations determine the necessity of what the spaces will turn into. Accordingly, the duty of whether this time would transform into one thing or another, or whether there would be any transformation stages at all, is undertaken by the role of time. All relevant places carry infinite knowledge, without which there would be no mobility. There is always a desire or things desired in the knowledge found in these places. Knowledge is achieved depending on the potential of desires in these desired things. This knowledge is the transformation into matter, a wave or a light. Likewise, the potential of what is desired depicts the magnitude of this expected transformation. This topic becomes more concrete when compared with the state of how poetry is written. In order to be able to travel in time, the physical state of the Prophet Khidr must have been transformed into a spiritual dimension so that he can adapt to change. The verb *return* is mentioned in the verses of Qur'an in this context. The phrases *returning for Allah* and *returning to Allah* can be given as examples. The Prophet Khidr's skill has to do with the dominance of these transformations. Thus, Allah's commands are delivered to the senses by entering the world of sensation; each of these feelings reflects time falling into space as much as the visible and invisible fate of every human being.

If we take this transformation as a means of help, as a standard, Allah does not need to be transformed; He is exempt from it. The fact that *Allah is closer to His servants than their jugular vein (Qaf 50:16)* is the declaration of great wisdom. The story of the Prophet Khidr can be summarized as follows: Escaping external factors, he was able to go beyond the values of dualism internally and set out after truth by the will of Allah. This virtue, knowledge and understanding had already been given to him from the very beginning. This situation brings to mind the Prophet Muhammad's ascent to Heaven. They are both honorable worshipers of Allah and representatives of the truth. They get their duties from Allah, and their purpose is to deliver the task taken over and cultivated in the universe to Allah. While the Prophet Khidr is on the horizontal line working for it, the Prophet Muhammad lives on this line and creates the ummah, where he strengthens the vertical line to reach the truth in all dimensions experienced on the horizontal line. In other words, he lives to reveal the gathered particles that make up the Fatiha. The Prophet Khidr thus knew the inner workings of time and shared what he knew with those who wanted to know. As the Prophet

Muhammad was present most profoundly within the *horn*, Allah called him to Himself. In return, Muhammad accepted Allah's invitation and the call to prayer, uttered by Allah, and then, he first visited the Masjid al-Aqsa and the Masjid Haram. Later, he rose to Allah to meet Him. The recollection of this inner working of the heart only depended on contemplation and trust in Allah. Thus, it emerged through the sensing of all the names and titles of Allah in this noble individual.

As with all servants who are participating in jihad for the sake of Allah, one person who was more important in the horizontal dimension was the Prophet Dhul-Qarnayn. He forms causes and represents mind. To end our topic with a story from the Surah Al-Kahf, the Prophet Moses visited the Prophet Khidr asking him if he could benefit from the Prophet Khidr's knowledge. As the Prophet Khidr knew the difficulties the Prophet Moses had in comprehending his activities between different times, he warned him that he would not be able to bear to be with him. Since the Prophet Moses had chosen someone like the Prophet Khidr as his companion, only then could the Prophet Khidr request of him not to ask any more questions regarding the incidents

2.2.9 THE TIME AND PLACE TO ASK QUESTIONS

On one hand, Allah wants man to consider everything that has been created. *Who remember Allah while standing or sitting or [lying] on their sides and give thought to the creation of the heavens and the earth, [saying], "Our Lord, You did not create this aimlessly; exalted are You [above such a thing]; then protect us from the punishment of the Fire." (Ali Imran 3:191)*

In order for human beings to confirm and research in this manner, Allah wants His servant to ask questions. He also highlights that the answers are hidden in the questions. This type of questioning corresponds to the fact that the knowledge of Allah is hidden within things. The knowledge in things is the way that leads to Allah. But here, Allah reminds us that everyone should behave based on their temperament and should not take on anything that is too heavy for their disposition, namely, essence to carry. Instead, if one has faith in Allah in this life of good and evil, always keeping Allah in mind and being decisive in trusting in Him will strengthen his understanding of patience. This is a precondition for being able to move and flow without being affected by the trajectory of time.

2.2.10. SIMILARITIES BETWEEN THE PROPHET DHUL-QARNAYN AND THE PROPHET SOLOMON

It is believed that the Prophet Dhul-Qarnayn and the Prophet Khidr were cousins. The Prophet Khidr represented the temporal teachings and

Dhul-Qarnayn represented the spatial teachings. During a particular battle, Dhul-Qarnayn appointed his aunt's son, the Prophet Khidr, as the vizier. This means that every form of jihad in the path of Allah is gained only by the knowledge of time and place. Allah gave power to both the Prophet Dhul-Qarnayn and the Prophet Solomon. Neither of them actually had a tribe. Yet, Allah gave the jinn and the devil, divers and animals into the service of the Prophet Solomon. He was a prophet who knew the language of all. The Prophet Dhul-Qarnayn had the knowledge of the cause and reason for everything. Both of them were endowed with measureless life wisdom. The Prophet Solomon's measureless sustenance was given to him after his flawless repentance, while the Prophet Dhul-Qarnayn's mind was his livelihood. In other words, Allah gave names and attributes into the prophet's service through His authority. He gave the prophet the authority to punish the people where he went or to invite them into goodness and beauty. However, on the path of Allah, the Prophet Dhul-Qarnayn chose what is most important to Allah. He separated the believers from the non-believers and preferred to chasten people. When Dhul-Qarnayn reached the valley between two mountains, he met a tribe *that didn't understand a single word*. While these were the ones who could not use their minds and were in need of the ability to use it, how was it that these people could trust Dhul-Qarnayn in such a short time? They asked him for help against Gog and Magog. Although initially they did not understand a single word, the tribe soon became aware of the beauty of the faith. They wanted to get rid of those monsters who prevented them from understanding and commanded them to commit evil. As long as Gog and Magog, the parents of evil, were not confined, their children would of course be the same.

The Prophet Dhul-Qarnayn responded to the assistance requested from him. According to the knowledge that Allah had taught him of how to take evil captive, he asked the tribe for iron and copper to make the necessary materials. When we look at the Surah Al-Hadid, we can understand that Allah sent down iron to define faith. This means that evil cannot take hold of places where faith is strong, but rather it is buried in the place it deserves. The tribe wanted to pay the Prophet Dhul-Qarnayn in return for his work. However, the Prophet Dhul-Qarnayn rejected this stating that any service to Allah could not be paid for. This is because beyond his awareness of the wisdom Allah had granted him, the Prophet Dhul-Qarnayn had devoted himself to all ways of wisdom. What could have been better than this wisdom? At the same time, the prophet stated that he had purely and simply fulfilled the duty given to him by Allah. He informed the tribe that Allah will destroy everything when His promise comes.

2.2.11 WRITING AND FATE

Writing is the impressive projection of a dimension of Allah and of the Qur'an, the word of Allah. In the thinking system of Brethren of Purity, one of the representatives of ancient Islamic philosophy, had been constructed as a macro and micro cosmos in the universe. This situation is actually supported by atoms in terms of physics. If so, isn't the *universe* also a *body* in the time-space continuum? Since everything in the universe is alive, wouldn't the difference be in the atomic dimension?

Isn't it this that portrays the necessity of the Prophet Muhammad and all other prophets to be human (this is also mentioned a lot in the Qur'an)? That is, isn't the soul itself energy? Isn't it the spirit that provides the functionality of energy to complete the mind's task? Of course, others are also a different kind of energy. Now, where exactly is the divine *light*? Wouldn't it have to be in the soul, the mind, and the spirit? If we were to begin from the provision that the existence of anything depends on the things that constitute it, where do you think these three are positioned? If the universe consists of atoms, should not these three be constituted of atoms, namely in one body?

1. If fate is a result of a cause-effect relationship, is it a matter of preference that human beings bring the lines of fate embedded within themselves into their lives (temporary causes and effects)? If so, then can human beings postpone or delay their death due to their preferences?

2. Wasn't the casting out of Adam and Eve from heaven caused by a reason or reasons that existed within themselves even while still in heaven?

Why was Satan able to deceive them? In short, can it be the existence of deceit and the ambition that results from or joins in this deceit, briefly could the nafs be found even in heaven? Is this true? If so, did Adam have to be a human being just to breathe in heaven? Why would his choice cause him to be sent to the earth? Was this supposed to be his choice to gain moral values on earth and return to heaven again? (Al-Hadid 57:21-22), and if so, what caused the latter?

2.2.12 DEPENDENCE-SERVICE

Be patient over what they say and remember Our servant, David, the possessor of strength; indeed, he was one who repeatedly turned back [to Allah]. Indeed, We subjected the mountains [to praise] with him, exalting [Allah] in the [late] afternoon and [after] sunrise. And the birds were assembled, all with him repeating [praises]. And We strengthened his kingdom and gave him wisdom and discernment in speech. (Sad 38:17-20)

As in all things, among people from the past, present, and future, names and attributes of the exalted Allah have numerators and denominators in the dimensions of all kinds of hierarchical geometric shapes of the concept of neediness that is abstract and vivid. Who is it that touches the deepest dimensions of these numerators and denominators? Aren't the prophets first in this? In other words, aren't those who are in need of Allah the most the prophets, caliphs, leaders and believers? This is because servitude and its mission are increasing to the extent of this dependency. For this reason, these noble individuals, who are *helpers* of the exalted Allah, are in need of his assistance the most during jihad, emigration and simple breathing. These helpers are the angels, mountains, books, birds, and trees of which Allah created by saying, *Be!* Even the jinn and devils working under the command of the Prophet Solomon...

[Remember] when you asked help of your Lord, and He answered you, "Indeed, I will reinforce you with a thousand from the angels, following one another." And Allah made it not but good tidings and so that your hearts would be assured thereby. And victory is not but from Allah. Indeed, Allah is Exalted in Might and Wise. (Al-Anfal 8:9, 10)

These verses emphasize that on one hand help occurs formally, and on the other hand, the power of Allah's names, such as the Forbearing, the Magnificent, the Ever-Forgiving, and the Indulgent are sufficient. May we, as humble servants, become in need of Allah more and more, and may His help and his helpers never leave us...

2.2.13 THEY DO NOT BELIEVE

The actualization of Allah's statement that *they do not believe*, or the occurrence of this phrase stems from the disregarding of what is right. Ignoring what is right has eventually a consequence. This consequence is, where there is no righteousness, deviance is prevalent. Allah states in many parts of the Qur'an that He already knows who has faith and who does not due to the weak temperament of human being. As in all the books, the Qur'an also states that in the mind of those who are able to touch true knowledge, events and those who pass through various events will be clear and come alive, that is, find life.

In another surah, Allah explains the reasons for the destruction of the tribes 'Ad, Thamud and others after explaining these stories to all people. By getting rid of the space congestion- the veils over the heart- He told a secret, just like those people who have visited places being destroyed every morning and evening... There are two issues to point out here:

1. Those who do not have faith will do it one day, therefore;

2. Opportunities and paths aimed at the attitudes and behaviors necessary for faithfulness and the strengthening of the faith are constantly presented (words and people walking between words...).

However, the starting point of faith is to think and ask questions about what is seen. The great book mentions that every wuestion has an absolute answer. Thinking upon what is seen is also a type of reading. The first surah to descend was the Surah Al-Alaq, which also emphasized it. This is because everything begins with reading. To illustrate: Who created human being? Isn't a fact that human being is mortal? If there is death, why not there be a resurrection? Thus, is it the things that are *rivals to Allah*, namely, the worshiping attributed to the space-time continuum, or Allah that leads us to absolute Truth?

Therefore, the more there are *rivals to Allah* in the way of life, the more insufficient the awareness of the right path. Therefore, it becomes impossible to find the right path. Instead, there is deference to doubt. However, doubt cannot substitute for anything. It can turn out to be a habit or even an obsession when people deceive themselves with the things happening around them or rather a presupposed world. Thus, man always persecutes himself, not others. The difference between the ones who see Allah and those who do not becomes evident here. People who prefer such a way of life may not feel that they pity themselves all that much. But after they die and are resurrected, cognizance of their experiences reaches them. They have a feeling that they have lived but for a particular hour of a particular day. After all, the rest is meaningless.

Here, life and the moment that is associated with life, namely, the question of what is truth comes up. Even tough the duo of life and moment can be thought of as living, dying, gaining possessions and children, the moment is within this. If life is taken to be simply living, then the moment is life itself. When life is lived as life, it turns into walking hand in hand with the beloved one catching each other's eye during that precious moment. Although it is painful and wistful to move away from this life, loneliness prepares the way to be reunited with the beloved one. To the unknown, yet safe paths...

2.2.14 AND THEY DENIED AND FOLLOWED THEIR INCLINATIONS. BUT FOR EVERY MATTER IS A [TIME OF] SETTLEMENT. (Al-Qamar 54:3)

Whichever quality of ourselves we strengthen, that quality becomes more and more accessible, so we become aware of it. Even if we don't see it, we can use its various particles as incorporators elsewhere. Because they seem to be achievable, they are permanent, and the duration of their permanency indicates their depth. Allah hasn't created any of His servants without intelligence, not really one. However, He has given them the responsibility to perceive, use and develop their minds. Everything has been given to them in advance. What man needs to do is to see the atoms of the mind, which is specific to human beings, and to process mentally and unite them, so that he may be guided anywhere he desires by the will of Allah. Everyone has their own path, whether this way or that. But there is always a path. On the way to that path, they need to use the appropriate signs (that is, paths) to reach that path. Everything is as true or false as it truly is. It is a matter of how you approach to the phenomenon and shape it.

2.3 ALLAH'S BOOKS AND THE HUMAN BOOK

2.3.1 CHANGE

...Verily never will Allah change the condition of a people until they change what is in themselves... To Him is due the true prayer: any others that they call upon besides Him hear them no more than if they were to stretch forth their hands for water to reach their mouths but it reaches them not: for the prayer of those without Faith is nothing but vain prayer... (Ar-Ra'd 13:11, 14)

In this example, Allah uttered in His noble book that it is absolutely necessary for the servant to ask of Him and strive for what he wants. In other words, when the servant asks, Allah creates the reasons, and the important thing is that like in the case of Dhul-Qarnayn, the servant needs to proceed towards these reasons to reach the desired.

Allah does not change them as long as people do not change their own faults or the attitudes and behaviors, which continuously obstruct them. In fact, the matter is the lack of hope and contemplation of Allah within these attitudes and behaviors, rather than changing them. Other human aspects in human being should also be kept in mind, for the human is cruel, arrogant and spoiled. He wants Allah to fulfill his wishes. Of course, in order for his request to be accepted, first, he must change. His prayer is already incomplete in terms of these dominant characteristics of him. In other words, human beings must first change in the right direction. However, according to Allah, if one person is to become more rebellious, he is given what it is that he asks for. Allah leaves this person's account to the hereafter. This servant has been abandoned to the worst and most unchangeable road there is. He has been held off from the mercy of Allah. They don't even know how to bring water up to their mouths. They have had such fallacies assuming that the things they have worshiped and idolized will help them or give them the knowledge required.

2.3.2 THE SYMPTOMS OF TAKING THE UNIVERSE BACK TO ITS PREVIOUS FORM

Diminishing from the universe: *...For each period is an appointment... Allah doth blot out or confirm what He pleaseth: with Him is the Mother of the Book. See they not that We gradually reduce the land from its outlying borders? (Where) Allah* *commands, there is none to put back His command: and He is swift in calling to account. (Ar-Ra'd 13:38, 39, 41)*

There is scientific evidence even if people do not believe it. The Qur'an, which was written 1,500 years ago, speaks of the erosion of today's real part. This is not only a source to learn from, it is also an indicator that the other verses are the truth themselves. Diminution of the ends of the earth by Allah is a sign of the apocalypse.

Allah points out that He will turn the universe back into water. This is because in the noble book, it is written that the universe was created on top of water. *And it is He who created the heavens and the earth in six days - and His Throne had been upon water - that He might test you as to which of you is best in deed. But if you say, "Indeed, you are resurrected after death," those who disbelieve will surely say, "This is not but obvious magic." (Hud 11:7)*

What Allah promises is to take everything back to its original form when He first started creating. On one hand, Allah tells us that He will restore everything, and on the other hand, He informs us that His good servants will be the heirs of the earth. The knowledge regarding what the earth could be like in comparison to the previous state is found in the expressions of physicists regarding the progressive creation of the universe.

2.3.3 THAT WHICH IS DECREASED FROM THE SERVANT

If the servant does not have faith, Allah reduces his tasks. Allah grants the servant other deficiencies in his work and strength pertaining to the strength and proportion of his faith. The important thing is the awareness of the asymmetric life he lives and his own mistakes in life he has comprehended, that he is willing to correct his mistakes, and prevent his failures from turning out to be a cliff.

Although deficiencies are conceived as both material and spiritual, they are solely spiritual in this regard, as the purpose in everything is to overcome the tangible and reach spirituality. Human beings, however, have little tolerance for material deficiencies. In this manner, the worst thing possible is for human beings to not remember Allah while they live in abundance and health, and are thus swept away due to their spiritual deficiencies. Worshiping the material corresponds to the term Taghut, which is the reason to seduce people and give birth to evil. In other words, Allah has created all things according to a specific measure, and these measures open paths that are not closed to the prohibitions and thus invite us to sin. Allah also gives opportunities to those who are in pursuit of this depravedness to multiply their depravity. Salvation is possible by avoiding taghut, namely *the open path*, as Allah puts it (see the section *Created with Righteousness*). One of the reasons for the prophets being sent into the world is to inform the ummah to stay clear of taghut. Deficiencies are also a matter of close concern for the dead, because they also have deficiencies. Allah knows what it is that the earth takes from the dead. This information is hidden, that is, it is preserved in the al-Lauh al-Mahfuz. The al-Lauh al-Mahfuz is the origin of all the books that Allah has sent down. When the servant passes away, what does he lose and why does he lose it? Moreover, what is lost does not completely disappear; everything is written in the book of Allah. *"But, [on the contrary], We have provided good things for these [disbelievers] and their fathers until life was prolonged for them. Then do they not see that We set upon the land, reducing it from its borders? So it is they who will overcome?* (Al-Anbiya' 21:44)

Here the word *we* refers to the noble names and attributes of Allah as well as of the angels. In other words, Allah reduces the terrain to a necessary extent according to other differences and variables through his names *the Utterly Just, the Magnificent* and *the Manifest*. It should not be forgotten that the life of the human is a terrain as much as the universe is, and the diminishment of this terrain is caused by gravity.

2.3.4 WATER

Allah knows what the earth takes from non-believers and what they lose when they die. There is a book that preserves this knowledge through the names and attributes of Allah. Instability of water indicates that it has the very essence of time within itself. Ninety percent of the structure of the universe consists of water. The human being was created from unstable water, from just an extract. We can list all the features of water as follows: water gives life to dead soil; and water becomes unstable unless it is combined with other atoms and compounds in human. In other words, the space-time relationship develops, and therefore a reduction occurs in the human due to the variability of the gravity. Age, sickness, the rejuvenation of the universe and all its objects are subject to this in different proportions. However, the same water also gives life to dead soil, conduces to sprouting from it, and it vivifies the human in the same way. In short, the human being comes into the world through inoculated water. After leaving the world and dying, water resurrects him just as water revives the soil after it dies. The resurrection of man will take place in much the same way as the expression that the two worlds are connected in a cycle.

2.3.5 SERVICE

And spend in the way of Allah and do not throw [yourselves] with your [own] hands into destruction [by refraining]. And do good; indeed, Allah loves the doers of good. (Al-Baqara 2:195)

When the concept of service is taken into consideration in a general sense, it means to act as part of a person's profession. A human being is a part of his profession and vice versa. In fact, the degree of respect and love for his profession while doing the work itself coalesces with that profession. This can be good or bad. If the person does not have the ability to meet the requirements, he may be unable to respond to the fundamental expectations of his profession and may end up not having the required strength. Or rather, by misunderstanding his profession, he may end up being dominated by it, which may lead him to artificial behaviors and attitudes. A teacher is at school during particular days and breathes in that environment's air. He becomes integrated with teaching methods. The atoms of his profession become dispersed as atoms and molecules throughout his own part of teacher universe. Receiving and giving in teachers and among teachers begin and end with similar situations and projections. A teacher teaches children with the very knowledge required by his profession. At the end of the effort shown by both the teacher and his students, at least the way for the children to learn this subject is formed. In the end, there is an observable event. The concept of service has a more specific and a more well-known aspect, which is it is to serve Allah. So, why did the prophets, who had understood Allah more than anyone else, face torture or were killed? Why was representing the truth so toilsome? The moment when the Prophet announced faith, it became a promise. The energy and projections left to the universe by the Prophet's posture and existence were in the form of very pure, clean and deep waves. The universe circumvents this promise because Allah has appeared here. Just like how the devil flees from Allah, so does one who is weak. He rejects the knowledge of the faith and mocks the person who serves upon Allah claiming him to be very dangerous. This is because the soft-tempered prophet continues to call people to renounce worshiping idols and their rituals. However, the hypocrite is so accustomed to his present life that he responds to the prophet based on his own social power by holding a negative attitude and thus negatively reacting to it. Persecution begins with the denial of the prophets. Allah punishes wrongdoers for the injustice they do to the prophets. *(Al-Isra'17:76: And indeed, they were about to drive you from the land to evict you there from. And then [when they do], they will not remain [there] after you, except for a little.)*

Allah destroyed these ruthless tribes. He sent the flood upon the people during the time of the Prophet Noah. The prophets Hud, Shuaib, Lot, and Saleh were also accused of falsehood and Allah finished off their tribes' depravity with calamities.

2.3.6 CREATED WITH RIGHTEOUSNESS

The tribes such as Thamud, Ad, Lot and Midian suffered calamities. The reasons for Allah sending these calamities can be explained by the damage they left on the main operation points necessary for the universe. This is because the earth, the sky and everything in between have been created with righteousness. If evil outweighs good among good and evil attitudes and behaviors on the human diagram, then those created with righteousness, that is, objects, feel the negativities as per the intenseness of this evil in their motion. However, because these things can be used or carried, they can either return the evil to the very person who is committing it or they can affect a certain segment of society, namely a few bad apples. Especially with the death of the innocent, their rights and sins are placed on the shoulders of those who do evil, as in Cain's killing of Abel...

Abel clearly said: *Indeed, I want you to obtain [thereby] my sin and your sin so you will be among the companions of the Fire. And that is the recompense of wrongdoers. (Al-Ma'ida 5:29).* The names Cain and Abel are at the very foundation. Even a single letter difference in a name makes a difference, and with that slight difference, history begins to write itself. From here, the infinite geometric angles and the uncertainty principle of intersecting points (Heisenberg) are apparent and wait for their return to the hereafter. In other words, the very emergence of a single angle occurs because everything actually begins in heaven. Adam and Eve are guilty of being deceived by Satan and are thus driven out of heaven. Adam was the first prophet. Cain and Abel were two sons born in the world after he was expelled from heaven. Cain carried the characteristics of his mother's and father's rebellious, envy, and endless desires, and thus he had much the same fate of being expelled from heaven. But would it be right to say the same thing with Abel? No, he carried the good parts of his mother and father in his genes and he was predestined to be a good believer.

Abel was always the one who watched for Allah's approval. There is a big contradiction here. Did Abel, who was such a good man, wish his brother to fall into a sin by killing him, instead of wishing that Allah would correct his brother? Was the reason for inciting him, the feeling of one who was living with the knowledge and love of Allah in opposition to infidelity or was it his own self-indulgent desire? Cain killed his little brother Abel and he was regretful of what he had done. As he did not know what he should do with his dead brother, Allah guided him. Cain learned what to do by seeing a crow scratching the soil. In this case, Allah also provided help to both brothers by allowing one to reunite with the earth and telling the other what to do. This story was the starting point of what humanity was expected to face. The seeds of good and evil were possessed by Adam and Eve in heaven, but they were planted on earth with Cain and Abel and not there.

In the Qur'an, Allah said that the cruel community or man in general was on *the open path*. Obviously, this was presented in two different ways:

a) The traces of those who succumbed to calamities are still visible in the world, and of course, these are found in people, too.

b) When *the open road* was considered in the sense that Dhul-Qarnayn took to the roads through reason, it appeared that the path meant the persecution of his soul rather than reason. In other words, while the tribes tended to ignore the same behaviors, homosexuality, injustice, miracles, etc., and when this became habitual behavior, these paths of desire were used. These paths were the ways to commit shirk and were open until death. This is the path, which does not require thinking about, and needs and consumes the most energy. This road is the flaming path of the devil. However, humans were created but from a drop of water...

The reasons that the object has been forced to hide at certain times are that it has been shaped negatively in opposition to evil and its negative variables in functionality since it was created with righteousness. The object created with truth is also affected by the evil in the universe, that is, evil causes objects to take a negative form, and thus there is a formlessness in the variables of its functionality based on the compositions of evil. Shapeless parts have their own accumulated special time (memory), unearthed in the object. When objects are unable to carry this accumulated charge in their own time, they appear in the universe as natural disasters caused by negative energy. When the heavy burden reaches a certain point in time, that is, has had its day, negative energy returns as a calamity. One thing that needs to be added here is the fact that these calamities have been marked by Allah. The foundational meaning behind the marked event is that the sum of those who gave rise to such culprits was already calculated, and the dimensions and the target of these calamities were already determined. What is unclear is that they can act simultaneously, including deviations in the mechanism of equilibrium, when these events concern different objects. Along with this, there is the sense of being able to construct the necessary components whatever the calamity should be and wherever it should reach. The situation that everyone accepts is that deviation is hidden within the very essence of the human beings due to their creation by Allah. So, humans can make mistakes, but the way to forgive these mistakes has already been demonstrated.

The wish to be forgiven by Allah is a sign of repentance and it requires that the same mistake not be repeated. In order for us to avoid mistakes, the seven verses and the Qur'an that had not yet been given to any tribes prior and that prevented people from being persecuted were given to Muslims by Allah. The seven verses make up, in general terms, the Fatiha. The term Fatiha is present in all the verses of the Qur'an.

Let's take division as an example. The numerator represents the good, the denominator represents the bad, and the middle line represents the truth and thus corresponds to balance. Mind that good and bad understanding changes places according to knowledge and intention. Then, the existence of knowledge is indispensable when it comes to the thickening of the line of truth. In a digital context, this line of truth varies according to the dimensions of its assets. When the unit's name is given to them, the line of truth for each unit is unchangeable, and its presence in the numerator and the denominator are thus directed, which results in gradual decrease of the effects of good and evil.

The angels have warned people through revelation that Allah created the earth and the heaven with truth. They also explain the reasons behind Allah creating with righteousness. We were created not to commit shirk but rather to seek His grace and to be blessed. Allah has offered everything in the service of man so that they shall benefit from it. The night, the day, the moon, and the stars all act by the command of Allah. One must go on quite a journey to seek for His grace. The paths on earth are represented by rivers and roads, and the stars are the paths in the heavens. However, some people still worship idols. The idols that are artificial and unable to create a fly are not even themselves alive because evil works are always dead. Thus, no matter how, whom or to what extent one worships other than Allah, since that worship is dead, the worshipped will find the same proportion of emptiness. The dead in this world do not even know when they will be resurrected. The reason they do not know this is due to them being distant from the truth. Thus, the buildings they build for themselves in this world have no foundation, because the objects of truth are not found in those buildings. They have no prayers nor can they perceive where punishment comes from. In the hereafter, Allah shall inform them through the Qur'an that the things they have mocked, namely the truth, will encompass them all.

If one does not touch the truth through his senses while living in this world, he will not be able to recognize it when he meets it in the hereafter. Moreover, just as a person feels the tension against the unknown and thus, experiences alienation, the same thing will happen in the hereafter. In fact, it will be even more severe.

People, who do not use their senses and ignore their heart by not believing Him, attribute their committing shirk to Allah's will. They defend themselves by making up excuses like *If Allah really wanted to, He would have lead us to the right path*. Being lead to the right path could have also taken place with miracles in part. However, Allah says that the duty of the prophets was only to announce. People who do not walk or think in faith, or know how to take water to their mouth don't change the way they believe even if they were to encounter miracles.

2.3.7 QUANTUM MECHANICS OR INNER (INVISIBLE) KNOWLEDGE

According to some physicists, quantum mechanics is a theory that needs to be based empirically. In their view, the results of experiments in quantum physics may be in contradiction with the findings of the pure soul, so much so that the pure soul seems to be a classical movement due to the absence of any conceptual dream. Although knowledge seems empirical, it is actually *history*. If today's date were March 12, 2015, life would include everything that happened up until this date. Otherwise, there would be a disconnection in the universe with this negative systemization, and disruptions in the orbit of objects would come up. It would end up with a breaking down of the various particles within the objects. In short, the order of the universe would be disrupted inadvertently.

The scientific explanation of knowledge being history suggests that the measure of knowledge depends on the quantitative multiplication of space-time. For this reason, knowledge is stable. Thus, it creates asymmetric behavior and variability throughout the dichotomy of space-time and/or further ambivalence due to its dualistic principle, and so establishes asymmetric points of connection as constants to divide the consistency and non-coherence in hierarchical decision-making. So, why do things and people need such a multiplication principle as a whole? Does not the need for proliferation make the system difficult to control? After all, why is not there a lack of control? Distress does not occur for two reasons:

1. The quantification of knowledge and its material,

2. Due to the material and mass of the universe—despite all kinds of chemical variations—it is always the same. For this very reason, knowledge is appropriate and materialized for use, and therefore, terminology, which clarifies certain things, has no incomplete imagination. In general, the true reality is that the unripe apple does not fall from the tree.

There are also examples that are far ahead of time and express exceptional situations. For example, there are some poets and painters who have only been fully understood after centuries.

2.3.8 THE INVISIBLE

They say: "Why is not an angel sent down to him?" If we did send down an angel, the matter would be settled at once, and no respite would be granted them. (Al-An'am 6:8)

There are some situations that have neither time nor place. Therefore, it is dangerous for man to ask for things that are not knowable, which can even result in death. This is because if the angels had been sent down to earth, then the structure of the earth would have to be different. The unseen would have to be visible. People, especially non-believers, would not be able to bear this. The reason here is the inadequacy of the touch of faith and knowledge being a disaster for man. Allah basically wants only one thing from His servant, namely faith. If the servant has faith in Allah, He forgives all the evil the servant has done. For Allah, compassion is a necessary duty. Almost all authority Allah has over His servant has two different meanings: to lead the servant to the straight path and to astonish him so that he continues in his depravity. At this stage, it results in Satan walking with the servant arm in arm. Just as Allah responds to the erroneous expectations of man regarding the things he wants and yet knows nothing about, Allah also warns the prophets not to perform miracles. Trying to please the non-believers and thus having pity on them would have been ignorant. The reason is that even if a thousand miracles happened to non-believers, they would still resist believing. Extraordinary situations do not bring people to faith. The miracle is the recognition of Allah's omnipresence, knowledge, and greatness. Sometimes this knowledge is not much quantitatively but may have overwhelmed the heart of the servant with faith in and love for Allah, almost resolutely. The Prophet Abraham set out to find the right path as did Dhul-Qarnayn. Before Allah sent his revelation, the Prophet Abraham believed in *tawhid*. He found Allah through a sound heart and good sense. The qualities of the Prophet Abraham are listed as follows: *Indeed, Abraham was a [comprehensive] leader, devoutly obedient to Allah, inclining toward truth, and he was not of those who associate others with Allah.* (An-Nahl 16:120) When he found Allah, he first let his father know and warned his father not to worship idols.

Allah ordered Prophet Muhammad to follow the religion of the Prophet Abraham, who represented these great qualities. There is a requirement for finding the right way, which is to believe and not to implicate injustice in faith. It is to cleanse oneself from evil especially the greatest evil, i.e., worshipping idols. The further a man keeps himself from evil, the higher his level of servitude will be. As a result, the one who seeks Allah finds Him. To do this, one must use the wisdom given to him by Allah. Allah has given man a heart, mind, intuition and various other characteristics. In terms of quantum physics, these are transmissive and transformation centers. When they are activated—thinking—the truth becomes visible or more likely to appear. As a result, if we look at the example of the Prophet Abraham, we see that when he abandoned the non-believers for the sake of Allah, He rewarded him. The coming of Ishmael, Isaac, and his grandson, the Prophet Jacob, were announced to the Prophet Abraham, who was over 90 years old. Allah exalted the Prophet Abraham for his faith, deriving Muslim believers from his generations. All prophets, including his sons and grandchildren, followed in his footsteps, namely clinging to his monotheistic religion.

2.3.9 THE PROPHET SOLOMON AND IMMACULATE REPENTANCE

The Prophet Solomon was a prophet who mentioned a lot about Allah and who as grateful; therefore, he knew a lot about Him. In exchange for this knowledge, Allah wants immaculate repentance. The beginning and the form of his repentance was related to the Prophet Solomon falling ill. He became like a corpse on his throne.

He understood repentance in such a way that when he got ill, he distanced himself from space and time. As he distanced himself, he repented and his repentance was immaculate. In other words, Allah's sight appeared in the Prophet Solomon's tired heart, where the good and evil disappeared. He offered abundance to the Prophet Solomon in return for his immaculate repentance giving him such sustenance that no one else could ever obtain it thereafter. This was such abundance that Allah did not expect any spiritual or material alms in return. In short, the pool mentioned is the Surah Sad and the words of Sad, which are the source of untold sustenance like the Tigris River, and it flows into its arms from here.

And if whatever trees upon the earth were pens and the sea [was ink], replenished thereafter by seven [more] seas, the words of Allah would not be exhausted. Indeed, Allah is exalted in Might and Wise. (Luqman 31:27)

The secret of sustenance can be handled with respect to the law of transforming matter into other states or waves as in the expressions of all the surahs more or less. These expressions depict the wisdom that can allow the source of innumerable sustenance as in the example of the Tigris River.

2.3.10 THE PROPHET SOLOMON AND THE SABEANS

The tribe of Sheba had been dominant for centuries. According to a rumor, Allah had sent more than ten prophets to this tribe and the Prophet Solomon was one of these. During the era of the Prophet Solomon, son of the Prophet David, who is thought to have lived between 970-931 B.C, the Queen Bilqis, the Queen of Sheba, was reigning. Bilqis conducted state affairs within a framework of certain solidarity. Even during that time, the main reason that the people of Sheba received such splendor was to call the society and their Queen Bilqis to faith in Allah.

2.3.11 THE PROPHET SOLOMON AND BILQIS, THE QUEEN OF SHEBA

Allah gave the Prophet Solomon much more sustenance except for the secret of the universe just like He previously gave to his father. Sustenance is given to the servant from birth to the extent that he can bear its responsibility and in accordance with the servant's nature. As for his other characteristics, the servant gains them through his achievements as long as he lives in this universe and Allah perfects him through the necessary words for the balance. This is because Allah does not give His servant a burden he cannot bear, and whatever the servant wants is always given to him. This occurs either immediately or as soon as possible. It is just important how much the servant knows what he wants. Thus, beneficialness of the desired thing and that its timing determined by Allah has not come yet are one of its decisive factors. In this manner, references are made from the present to future situations that the servant will experience, so that he can adapt to the conditions of his life and his strength increases.

Allah also gave many characteristics to the Prophet Solomon. However, just like the prophet of patience, the Prophet Job, first lived a magnificent life through the grace of Allah and then had his children and everything else taken from him (the dimensions of the characteristics and their purpose were different for the Prophet Job), the Prophet Solomon also lived through various difficulties. Before the Prophet Solomon met Queen Bilqis, the animal kingdom, jinn, devils, and various other creatures had been given over to his service. The Prophet Solomon, thus, became a prophet, a monarch, and a corpse on his throne by becoming ill when he was very rich. He wanted from Allah's mercy an eternal splendor that no one would be able to reach after him by virtue of his repentance. He wanted property to be even more in love with Allah's power and to be ever more grateful to Him. Allah already told the Prophet Solomon, *They made for him what he willed of elevated chambers, statues, bowls like reservoirs, and stationary kettles. [We said], "Work, O family of David, in gratitude." And few of My servants are grateful. (Saba' 34:13)*

The will of the Prophet Solomon was of course Allah's will. In other words, Allah wanted this to happen so that the prophet would be ready to face the situation awaiting him. And as it is known, the queen of Sheba people believed in Allah being influenced by such splendor. The fates of the Prophet Solomon and Queen Bilqis intersected at that point. When Bilqis entered the palace of the Prophet Solomon, she pulled her skirt up slightly so that it would not get wet. But she soon realized that she had been wrong. This magnificent, crystal clear palace touched Bilqis so deeply that she believed in Allah right there and felt regret for all of her faithless years. Bilqis's body could not move due to such beauty. According to these two valuable thoughts, her way of thinking and speed slowed down and became stagnant. This resulted in a reverse perspective: Bilqis's astonishment and admiration conduced her own soul and heart to look at each other (nucleon). Put it differently, reverse operation of atomic functionality leaves electrons in the vital line instead of the neutron holding electrons in motion or moving them differently. Thus, the electrons increase the proton's mobility and this time, the proton puts the electrons at its service. In other words, during the space-time paradigm (let's not get into the subject of quarks yet), the electrons begin to move. Then, the proton absorbs them and the information is transferred to the neutron. Neutrons accumulate what they receive and leave the rest to the universe allowing the electrons to react again.

2.3.12 HARMONY IN BODY AND SPIRIT

the external doesn't care what will happen
combining reasons for their own sake
not even itself knowing its problems
rushing around just for the external

the heart or perhaps the soul
cannot bear staying
as the external is different
being itself a stranger

the heart becomes sick
with a deep longing
asking the external to be different
this time they both weep

Without the soul, what's in the heart cannot flow through the veins. In a more general sense, it is the soul that helps the body to adapt to change. This is because every instance that an organism is under any particular influence, the soul is always in place to maintain its order in the good and evil in which it experiences and causes. The reason is that it is the very thing that has a spirit and provides harmony, and this spirit comes into existence through Allah's breathing on His servants to the extent that He considers it necessary. The phases in the work of the soul and organism regulate the velocities of senses i.e., electrons in particles in their orbits. According to this setting, an irregularity develops. Then, the infinite variability of the irregularity (small invisible entropies) and the order (visible entropies) spread as data throughout the dichotomy of the organism. In short, the substance is geometrically proportional to diagonal dimensions. Therefore, the finding of the diagonal includes the entire existence of the human and the universe, which is infinite according to the quantum mechanics approach. The only difference is that human beings tend to experience life rather than comprehend it. The result is that man lives only as much as he knows Allah, namely to the extent of the Fatiha of the Qur'an. As Allah said, *[Allah] will say, "How long did you remain on earth in number of years?" They will say, "We remained a day or part of a day; ask those who enumerate." He will say, "You stayed not but a little - if only you had known."* (Al-Mu'minoon 23:112-114) In this manner, the purpose and meaning of life is: Man lives only to the extent that he has fulfilled the duty assigned by Allah so as to complete the divine light. This may not make a huge difference in a lifetime between two completely different people in terms of their faith in Allah, as per the theory of relativity, which is the subject of space-time calculations. But in the realm of life, this occurs differently. Everyone is involved by the grace of Allah, but the real issue here is realizing that one is conducive to the occasion and obeying Allah's commandments and prohibitions in all their forms in this life. That is, his right to be an heir of his forefather is entitled by Allah and the servant's current deficiencies are compensated.

2.3.13 ALLAH DOES NOT HAVE SHORTCOMINGS

It was at the holy valley of Tuwa on the Sinai Peninsula of Egypt where the Prophet Moses met Allah for the first time and that place and time were blessed with hearing the voice and the glory of Allah's names and qualities. Allah, the Lord of the worlds, does not have any shortcomings. He manifests Himself only in places where His voice is utterly true. Those places are rendered holy to reveal Allah, the Sublimely Exalted and Magnificent.

Allah presents truth either one-to-one or indirectly. The truth presented here is one-to-one.

Where there is truth there is exemption from shortcomings. Allah is the truth and very manifestation of truth. Therefore, shortcomings are compensated according to the size and intensity of the truth. In this context, there are miracles at the very time and place of truth. In the place where there is cruelty and pride, miracles are rejected regardless of whether or not they are known or seen.

Speaking about the science that was given to the prophets David and Solomon, the importance of science is underlined confirming the previous verses. The importance of truth and reaching truth is stressed. In this example, where there is science, nature's truth becomes apparent. When they are apparent, truth ends up playing an auxiliary role, e.g., birds, ants, mountains, wind. *Say: "Praise be to Allah, and Peace on His servants whom He has chosen (for His Message). (Who) is better?- Allah or the false Allahs they associate (with Him)?" (an-Naml 27:59)*

All that Allah has created is good, as long as it does not deny its servanthood. For servants, everything becomes sustenance, blessing and hereafter. People's sustenance is reserved within the stars in the sky. The prophets cannot make themselves heard by the dead, and so it is necessary to hear, smell, and see before you die. Unfortunately, it is futile to make the deaf hear and make the blind see.

2.3.14 ALLAH'S BOOKS AND THE HUMAN BOOK

Every man's fate We have fastened on his own neck: On the Day of Judgment We shall bring out for him a scroll, which he will see spread open. (Al-Isra' 17:13) In other words, there is a book, in which everyone writes his own life. And this personal book indicates how much the servant puts—or doesn't put—into practice the verses of the Qur'an that give advice, commands, and words to obey. By having his own book, no one will be able to take on another's sins. However, Allah shall not pour out his wrath upon the earth unless he first sends a prophet to warn the people. The number of books that have been written, and those that are going to be written, is equal to the number of these books. All books are present within the al-Lauh al-Mahfuz, the holy book, which is next to Allah. The al-Lauh al-Mahfuz is a book, which includes the names and attributes of Allah, infinite knowledge in terms of singularity and plurality, and thus, the fact that this knowledge covers the seven layers of earth and sky. *And if whatever trees upon the earth were pens and the sea [was ink], replenished thereafter by seven [more] seas, the words of Allah would not be exhausted... (Luqman 31:27)*

Every human being is responsible for their deeds and there are two near and far reasons for this. Allah's verses are on the horizon and within the man's soul. The soul is mobility, which determines the proportion of balance consisting good and evil, and the faith formed by good and evil. In both cases, there is an artificial balance or a balance of truth. As a result, the soul ends up having a systemic structure and this structure reveals how the structure of a person works, that is, the truth. Thus, Allah is a witness to everything here. The fact that the same verses are on the horizon is explained interdependently as in everything else. The first explanation is that man sees what is far away, not what is near. The other explanation is that human beings leave to and receive from the universe what they have done. The last explanation is that as long as variability in the virtue of differing purpose according to the order of Allah is taken a criterion, in situations when his servant is astonished, either the astonishment happens so that deviations increase or help is obtained through revelation from the universe and/or the descent of angels.

The issue that everyone should know is that every situation carries knowledge within itself that reaches to Allah. Allah's names *the All Knowing* and *the All Aware* point to this. This is an indicator that Allah encompasses everything. If Allah gives a specific task to His servants, He will make them serve their people with His *words*. But those who have evil habits will be kept away from Allah's verses and revelations because they accompany Satan. When all angels prostrated before Adam, Satan refused. Because the event took place in the council of angels, Allah never forgave Satan. The reason Satan didn't prostrate before Adam was that he considered himself superior, that is, his pride. He felt being created from fire was superior. Allah commanded the devil to descend from his position. Satan promised to become an obstacle in front of people on the righteous path. Therefore, it is only a matter of time before the person on the righteous path goes astray. In other words, in whatever area the faith is weak, Satan attacks it in particular. He makes the way for all opportunity for all kinds of wickedness, and keeps the worshipers out of gratefulness, just like himself. What needs to be understood is that everyone is human and no one is perfect. The important thing here is that there is a balance between good and bad dualism, which is established through the righteous path between deviation and non-deviation.

There is also balance lest the deviations increase, but this balance is false and devoid of faith. However, according to a situation in the al-Lauh al-Mahfuz, in the spirit world, Allah took their words as a promise that they will not say anything else. They've already read the book. As everyone knows, promises require accountability. Being good and fair brings about good results. There won't be any deviation from the target or else the deviation will decrease according to the intensity of the good.

If there is no revelation to a person, he has no knowledge of anything. That is to say, to follow what he does not know means to hold the ear, the eye and the heart responsible. Man should know his own limits. Verses are reasons that lead to eternal purposes. Some of these verses that have descended for a thousand various reasons serve to scare people. One of the symbols of fear is the cursed tree mentioned in the Qur'an. Why do people talk about such a tree to create fear? What does the tree represent? For whatever reason, the common purpose of all is to strengthen faith. For this, they all are grace offered to people by Allah. Therefore, the meaning of the verses is that they are miracles. To deny Allah or the divine power in the name of Allah leads man to a negative world having been deprived of man's supreme power.

Believers and scholars are quite afraid of miracles, because it is not evidence or exaltation that is requested by the believers. What they want is only to attain to the mercy of Allah. But most people have no intention of having faith in spite of the verses. Evil seems to flow in their veins. If Allah abolishes the verses, which He has revealed to His messengers and servants, the protectors provided by Allah will be lost. In other words, if there is no verse for the servant, there is no situation either. This is binding for both the hereafter and this world.

2.4 THE SYMMETRIC SYSTEM OR ALLAH'S ASSISTANTS

2.4.1 THE SPACE-TIME CONTINUUM

There is an important sequence in the first four verses of the Surah Al-Rahman. First, the Qur'an is taught. Then, the human body is created (servant = soul and human = body?). Finally, Allah teaches man to explain the Qur'an. Thus, it is understood that a presence and balance preexisted before coming into world. Allah, of course, created the earth in a balance and order.

However, the seven-level earth and sky were created after the creation of man in the spiritual world and the hereafter according to man's particular characteristics. The reason for this was to create harmony between them. When the universe was created, the verses and their units that were to be in the space-time continuum were processed into the universe. While some of the necessary measurements were visible, there were some invisible parts. Both would make their own independent order by distinguishing themselves from both visible and invisible dualism according to the level of knowledge in the future.

Here is another way of looking at it: The universe is in harmony with the child, the adult, and the place and time in which they live. Even the natural disasters of the universe are in harmony. The universe cannot remain indifferent to mobility, i.e., it reacts to the act of worship and being worshiped. This is because if everything is built with an order, there should be a measure and a number of this measure. Because the parts missing the truth distort the balance in the space-time continuum of the universe within this order, the universe, which constitutes and accumulates some negative redundancies up until the time the apocalypse breaks forth, finds a solution by throwing out the excesses carried within or bursting out to get rid of them. This is to say that the continuum of the universe, i.e., those who were created, also exists in the hereafter according to the principle of space-place. Also, the clearer each man puts forward his own eternity, the clearer it actually becomes his eternity or afterlife.

Of Him seeks (its need) every creature in the heavens and on earth: every day in (new) Splendor doth He (shine)! (Ar-Rahman 55:29) What does it mean *to want*? At its very basics, revolving in an orbit, consciously or unconsciously (active and passive) for the sake of Allah, is already a state of wanting. The stars, the moon, the people, and the jinn return to Allah's will with gratitude. The important thing is to see as much as possible where and for whom one revolves.

2.4.2 PEOPLE AND REPRESENTATIVES

One of the most important scientific events of our day was the blasting off a spacecraft, separation of its module, and the module's successful landing on a comet ten years ago. However, the life of that module was not very long. It lost its energy within three days and shut down, as it did not have the ability to recharge. These type of initiatives are not the starting point, path, or destination of the item inventions, but rather they are the revelation of what is present in human beings. All information is given to man by Allah. The emergence of some information is related to how much the person carries that information in himself; in other words, it is related to and in proportion to how much it is combined with other information.

To illustrate this, there are units of the earth and the sky. These are the flow of knowledge offered by Allah. All units of the universe created by Him have the same or at least a single dimension in this flow. This dimension is based on human knowledge, and thus the illusion of indefinite dimensions prevails. On the other hand, with such a system that appears to be negative, it is only possible to reach a conclusion that takes on a singular knowledge, and this is provided by the overlapping in the flow. This is because Allah takes His servant closer to Him in proportion to his knowledge; and He allows his servant to breathe in and out with the intensity of the names and attributes that He chooses for His servant in line with that knowledge.

Therefore, these units are the same in every person in terms of characteristics, but they have their own geometric shape and intensity according to the knowledge of each one of us. It becomes inevitable over time for this unit, which grows with its way of life, attitudes and behaviors, that is, growing in its own structure, to be in touch with its other structures.

To give a concrete example, the sky is used as a type of laboratory as people send up gadgets. Of course, laboratory environments are needed to verify (hypo)theses. But the main emphasis is to find a unit of human beings that understands the meaning of searching for something. Thus, the field and application of this investment is human beings themselves, and their return would also be to themselves and to others.

How can one be certain that the Pleiades exists if one does not know that it is in the sky? Here is a question in the Qur'an that Allah has emphasized regarding the consequences of killing a man: Why does it seem that a man has killed all men if he has only killed one? Could there perhaps be two answers to this question?

1. Since all people now have to carry the unit of the deceased, their burden ends up getting somewhat heavier.

2. Because of the first reason, all people who are obliged to carry this burden are differentiated by their apparent characteristics, partly by those who are involuntarily burdened. As a result, they may cause disorders in the organism as well as lead to a brilliant attitude and behavior. If the characteristics of individuals are infinitely variable, then measuring people's intelligence (I don't think there is intelligence, but there is only potential) can only be possible by evaluating variables that can take on more than one value, as well as through evaluating them according to the number of these variables.

In fact, the leap the education system took in 1983 through psychologist Howard Gardner's seven criteria and the development of his theory of multiple intelligence consisting nine/ten distinct abilities. Some of the measurements cover topics such as the existence of scholars, different developmental continuities, and the abilities to meet them. In his research, Gardner explained the criteria and skills in line with intelligence. However, it would be more accurate to consider that intelligence is a very complex and flexible structure. Then, it would be highly objectionable to explain this complex structure by associating it with the concept of intelligence alone, leading to the ignoring of the very virtues that Allah has given to all people. This is because Allah created all people according to a specific essence, and He blew His soul into that essence. Thus, how can this potential and those who possess it be deprived of His word? What if all people were created just for the parts of that word to be revealed, whether willingly or not?

2.4.3 THE FIRST JIHAD IN THE HEART OF HUMAN AGAINST SATAN

The Surah Hajj is the surah that deals with being on the righteous way (as-sirat al-mustaqim). People and the prophets are human beings. Therefore, Satan may try to put in space and time based human desires, whether good or bad. When other desires are added into a single and pure desire, the pure desire disappears among them as it is pushed backwards. In the end, that pure desire emerges in the heart again as the place of direction and rotation.

Nevertheless, Allah can block Satan's path here. Where there is faith, it becomes even more obvious to see Allah and turn towards Him. Then, Allah puts His own verses solidly in the hearts of men. Serving upon Allah is set in the hearts of the prophets as truth. This means that being independent from time and space is possible only when the heart is in the same direction as the truth. Just like other people, the prophets also struggle against Satan and nafs. This war can only be won through knowledge.

For two reasons, Allah allows Satan to tempt men:

1. To wake them up and find out whether or not the ill-hearted and cold-hearted people would respond to His call.

2. To increase faith in Him and attain peace and satisfaction in their hearts of those who are aware that the knowledge given to them is the very truth that has come from Allah.

Allah undoubtedly guides believers straight to the right path. Those who are given knowledge will immediately recognize what the devil has done. They reduce the influence of Satan by distancing themselves from him and even neutralize his effect. This is only possible with the tendency of the servant to Allah and Allah's support to him in return. Allah thus guides those who are given knowledge to the right path. Otherwise, the ill hearted distance themselves from the truth and thus, allow Satan to settle in the orbits within their hearts. They are pleased by that. However, humans have to obey Allah's commands to conquer the devil. Shaping their behaviors according to these commands is to know one's owner and his helpers. This is how the servant's right to cling to Allah Almighty realizes.

2.4.4. THE PROPHET MUHAMMAD AND JIHAD

The life purpose of the Prophet Muhammad and his believers is jihad in the way of Allah. During his jihad, some believers behaved reluctantly and cowardly. Thus, the Prophet Muhammad asked for help from Allah, and Allah accepted his prayer and supported Muhammad's army with His angels. But first, Allah put the believers into a light sleep to give them confidence. Then, He poured water from the sky over them. This was to cleanse the believers and remove the filth that Satan had left on them. In this way, their hearts were connected to each other and persisted throughout the war.

2.4.5 THE HEGIRA

The hegira is to give water to a thirsty flower, to feed the soil if it is lacking minerals in order to supplement the plant with its necessary minerals, but at the same time, to know that these measures will not protect the flower from extreme external factors such as cold and heat. The hegira is changing one's path to avoid stepping on an ant, knowing the creator of that ant and thus not hurting it. The hegira is to walk in the name of humanity and know when to stand for it. Stopping, on the other hand, is to stand against what you defend for the love of mankind and not letting anyone get hurt.

2.4.6 THE MESSENGERS OF ALLAH AND THE HEGIRA

In the hereafter, property only belongs to Allah. There, the fight for property and idolization will be done away with, and those who refute the verses of Allah must know that the day of absolute punishment will come. The choice is left up to the servants. The hegira, in the way of Allah, is an option that the heart of human will be satisfied with. He always circumambulates his own Kaaba, by shrinking like shadows and turning to the right and left... The very form of the hegira is to treat all diseases of the heart and remove the veils, i.e., the enemy of knowledge, and thus, one establishes his own Kaaba within his own heart.

The other option is to deny Allah. This is the way of the hypocrites and oppressors who do not even notice the heartbeat. They are deceived by Satan because they deprive themselves of the knowledge of faith and accompany their artificial world through serving him. To increase their numbers, they endeavor to destroy the heirs of beauty and faith; and in order to be successful, they, mostly persecute those people who are influential. Foremost among these are the prophets, the leaders and the scholars.

Many prophets and scholars were killed just because they had served upon Allah. The lives of those who were not killed were also subjected to persecution, suffering and struggle. Allah will ask for an account of those who are wronged and His help will support them. Allah will bring into being the words and support that He has set in night and day within the whole universe. The touches of faith are already a means for these lights, which will eventually come to life through the command of Allah and the help of the angels. If a person is wronged, it is his right to react justly. In that, he may respond to the injustice he has faced by his own will and may intervene when he sees such a case. When he is wronged for a second time, if he cannot afford to intervene, he must remember to stand with Allah. Allah is able to help those who have been wronged. This is such a help that Allah will do all that is necessary for a single servant, and by doing fine-general calculations, the very act of help will be set in the universe for that particular person.

There is a fallibility concerning those who are unfair and who are victims, since in the end, we are all human beings. Therefore, abstaining, right thinking, and avoiding self-deception will be better in the pursuit of righteousness. Allah is the only one who hears and sees rightly. In some cases, there is no right to expect help from Him. Let's say that a person has much less right than he thinks regarding events or individuals. Then, it is decided by Allah to give the person this right when it reaches a necessary amount, such as when there is a lack of right or it's an area that belongs to another person.

2.4.7 THE SYMMETRIC SYSTEM OR THE HELPERS OF ALLAH

Of course, Allah is contented. He is the Indivisible and the Everlasting. He doesn't need anything, nothing at all. Allah is everywhere. He, who is the owner of earth and sky, has helpers. They are the very ones that try to accomplish His commands in such a way that He shall be content and pleased. Angels, people and things are as if in a race to exalt themselves. This knowledge of the numbers and that of who and what are involved in the task is kept by Allah. In that, the existence of the seven-layered hell and heaven, in terms of degrees, is explained concerning how much the servant clings to Allah's rope or the suppression of the nafs.

From the standpoint of the servant, being on the hegira and in jihad on the path of Allah means to hope to be one of His helpers by His will. Allah brought the seven-layered universe into being without a prior example. All examples come from Him and return to Him.

Just as Allah is the first teacher who taught the Prophet Adam all the objects' names, everything is a living teacher for the lives in the solar system. Even a teaspoon can remind us of things. Maybe it will show us things that did not notice at the time. This can always happen in different ways, though it is a repetition. Thus, this is a symmetric system. In this system, everything and everyone is responsible for each other, even if they are unaware of this. The particle is responsible for the atom, and the atom is responsible for the particle...

Gabriel (PBUH) is one of the helpers of Allah who created the universe out of nothing. However, Allah did not create him only; therefore, he cannot be the only helper of Allah. Let's just remember that he's one of His devotees. Therefore, Gabriel also has helpers. Gabriel's (PBUH) helpers are Michael (PBUH), Azrael (PBUH), Raphael (PBUH), as well as some other angels and people who were appointed as helpers. The opposite is also true. Helping as per the requirements of the system that works in the past and eterne systematically replaces with its whole geometric shape or the objects certainly share their gains. In other words, every object and living creature takes on their own responsibility according to the words and concepts they carry, i.e., their Fatihas, and leave them in the universe.

The Fatihas are everyone's own right that they take on directly through knowledge of the world. But when the Fatiha is considered as the truth of Allah, everyone grasps this truth to the extent that he comprehends it. It means that Allah's helpers must have multiple worlds based on the size of their duties. To be Allah's helper, Gabriel (PBUH) appeared to other angels, things and people when necessary based on Allah's command. If necessary, he becomes the system based on the very principle of integrity, that is, he will either descend into the water or, if necessary, will himself be the water from its center, the star Sirius, to the universe. Gabriel (PBUH) is one who is closest to Allah and one of Allah's most beloved worshipers, who is capable of being anything or anyone infinitely.

2.4.8 THE NAME, THE MOST GRACIOUS

The obligation of compassion, given by Allah, is hidden in the name, the Most Gracious, and again becomes apparent with Him. At the moments of compassion, the name, the Most Gracious, turns into the Most Merciful. Thus, the servant turns toward his Allah and qibla. The closedness of the name, the Most Gracious, as well as the name, the Patient, is to expect the call of His servant. This is because Allah, who is Right, gives all His servants the right to live. And He shall not deny what is in Him for His servants and the rights here. This is because these concepts cannot be separated. And there are some rights that have unchanging features within the name of the Most Gracious. Even though they carry different characteristics that differ from person to person, the right to live, die and resurrect are the rights given to all living things and even objects created by Allah. In addition, the right to eat and drink is one of the principles that does not change. However, those who pursue mischief would encroach on all necessary rights of others. While Allah serves man without end, what is offered back to Him is ingratitude and its fruit.

2.4.9 THE KAABA

Sacrificial animals have several benefits up until a specific period of time. After that, the place where they arrive is the Kaaba. That is to say, due to the sacrifice, a person makes both he and the people around him will have a benefit for a specific period of time. Then, this good deed will reach the Kaaba. The meaning of the sacrifice is the Kaaba, the destination of every beauty done for Allah and the heart of the universe. The macro heart connotes all the micro hearts that live and die for Allah and are seen worthy to be instrumental in adorning the Kaaba and its surroundings with divine light covered with another divine light.

The end of those who are not heir of the Kaaba is depicted in the Qur'an with a destruction they will suffer. The collapsed ruins point to memories as if the souls of the people who have passed away are still there. It is evident that living people should positively direct their lives by learning from these consequences. This is a warning not to waste the fleeting life. But the hypocrites' hearts are blind. The hypocrites who mock torment want it to come quickly. Allah will not back on His word of torment; torment will surely come upon them. One thousand years for humans is but a day for Allah. That single day flows from the seven-layered heavens to the earth, as creation begins. For this reason, the names of Allah, the Hidden, the Manifest, the Sublimely Exalted, the All

Subduer, the Most Gracious and the Most Merciful manifest. Allah's creation of man is symmetrical, i.e., perfect. While the universe has countless differences, their infinite visible and invisible combinations also represent how flawless human beings were created. However, the hypocrites are unable to read and understand themselves and the universe.

Another perfection of creation is the existence of a real book, which keeps the similarity of the attitudes and behaviors of all people that have existed, and unlike everything else, which is an unchangeable constant. It should be remembered that Allah has given a part of His soul to His servants.

On the other side, Allah created human beings from the essence of perishable water; thus, the human is asymmetric. This can be explained through two interdependent things. First of all, he is mortal. Secondly, because he is mortal, the need for creation arises out of a duality. He grows and dies by giving correct-wrong meanings to life according to value judgments. But these are already for the determination of faith. Those who conceive gratefulness within this inexpressible order as a misfortune are following the way of Satan.

The ones who built the foundations of the Kaaba were the Prophet Abraham and his son Prophet Ishmael. They built the Kaaba together to fulfill Allah's command. They obeyed Him and showed complete surrender. To fulfill Allah's commandment, the Prophet Abraham would sacrifice his son the Prophet Ishmael. Both father and son were being tested by giving up their own lives through the sacrificial dimension of the hegira; one was sacrificing his own son and the other sacrificing both himself and his father in the way of Allah. This is the Kaaba of surrender to Allah and/or the Kaaba stone of the hegira. Through this test of surrender, the Prophet Abraham's faith in Allah was tested. The reward for the difficulties encountered in the hegira before and after this, which will flow into time was given to the Prophet Abraham's descendants due to the fact that they were also messengers in charge of the invitation to Allah. There are other rewards concerning this. In this sense, the beginning of all was to build the Kaaba in the heart of the universe. Through the existence of the Kaaba, its particles take residence in the other believer servants' hearts and even flow through their veins. That is, the body of the human becomes the Kaaba. When this transformation takes place, the uniqueness of Allah and He Himself becomes more visible. Fulfilling Allah's commands to reach Him is a means of reaching the truth. Everyone's actions are the activities of the colors and dimensions of the Kaaba, both in his own heart and in parallel to it.

2.4.10 TO BE RETURNED TO ALLAH AND HIS NAMES AND TITLES

It is We Who will inherit the earth, and all beings thereon: to Us will they all be returned. (Maryam 19:40)

The servant serves Allah. He shows loyalty to Allah's commandments and becomes the Kaaba itself by shaping his life in such a manner. When he becomes the Kaaba, he is shown His path. He advances on this path through Allah's names and attributes. By being on this path, everything and everyone is influenced by it or directly takes it. They circumambulate around the truth, that is, the divine light, which appears in the servant. The one who circumambulates the closest to Allah is the person in whom Allah is clearly present. The simultaneous phenomenon that occurs here is the dissolving of the servant with righteousness as he first shapes himself through circumambulating according to the very effect of love and then by completely leaving general space and time. In other words, human beings can only get out of space and time when they dissolve in truth.

According to the extent and intensity of love, the servant also travels in and to Righteousness. Here, obedience to the Most Gracious Allah through the understanding of the servant's faith is underlined.

The right that the Most Gracious Allah has bestowed on His servant should not be forgotten. The servant lives with the particles of truth that are necessary for the world of motion (truth) and its activity (soul) until his last breath of life. It is he who repents and turns to Allah, and from those who repent, Allah destines his repentance to continue until it is perfect.

2.4.11 FORM AND CIRCUMAMBULATION

Humans exist as long as they circumambulate around the supreme beings to the extent of their knowledge and experiences, so that they exist. If Allah has made human beings His heirs and representatives of Him on the earth, then people, things, and the heirs of the hereafter, near and far in terms of their heirship, revolve around this being. For example, just as everything is required to revolve around the sun due to the order of the universe, everything and everyone that exists and has existed will also revolve around someone who is like the Sun to them.

2.4.12 IRON AND FORCE

There are direct (symmetrical) verses of Allah in iron. These verses help the asymmetric universe to maintain the mechanization of it and the formulations of symmetry of this mechanization emerge. Through all these signs, Allah's mercy and compassion can be found.

Who is he that will Loan to Allah a beautiful loan? For (Allah) will increase it manifold to his credit, and he will have (besides) a generous reward. (Al-Hadid 57:11)

Whoever handles the iron beautifully, his good words (= behaviors) reach to Allah and take that person to Him. *Indeed, the men who practice charity and the women who practice charity and [they who] have loaned Allah a goodly loan - it will be multiplied for them, and they will have a noble reward. (Al-Hadid 57:18)*

This is all kind of sustenance given by Allah. And there is much more! As the servant handles iron, oo determine the faith, which is the real permanent, his actions become divine light. This light illuminates, from his front and right side, his way to heaven through the darkness. These are those who are on the path of truth, who believe in Allah and His messenger. Allah bestows a light to follow only for those who are on the way to truth.

The order is first the faith, followed by the truth. This is because whose word is true other than the one who has faith? He becomes the Fatiha. And in that, the nucleons bend in harmony, circumambulating for Allah in front of him as well as his right. The Fatiha, the Prophet Muhammad and Gabriel ask forgiveness for the other servants who have received their share of the skills of these supreme beings. Those who have a place with their Lord are the right ones and they reach the rank of martyrdom. And, Allah says that *also he (the servant) has a very precious reward.* This is a reward that Allah keeps to Himself. It is a very different and very worthy one that human beings cannot conceive. People shouldn't think that they can do everything on their own. What people do is to realize verses that exist or have existed by the will of Allah (determinism), which are already present in the space-time continuum. The person who knows this will neither get upset when in trouble nor will he get spoiled (contempt-resignation).

No misfortune can happen on earth or in your souls but is recorded in a Book before We bring it into existence: That is truly easy for Allah: In order that ye may not despair over matters that pass you by, nor exult over favors bestowed upon you. For Allah loveth not any vainglorious boaster. (Al-Hadid 57:22, 23)

Iron has many different varieties. These certainly have different contributions to the universe and to people. For the determination of the faith, the combination of the iron that has descended and its components must be examined. Which faces of faith are present, which colors they have and why they are present in this combination should also be discovered. The path is the Fatiha, the direction is to be as close as possible to Allah.

2.4.13 THE IMPORTANCE OF AWARENESS

Allah calls the earth and the sky to Himself, regardless if they come willingly or not. The fact that both of them are willing and come depicts the very perfection of their surrender. The seven heavens and the earth were created in six days. To control them, Allah took the domination on Himself. The place of creation is not known.

However, the ninth heaven existed before the creation of the earth and the sky. There are angels who carry and encircle the ninth heaven. There are also believers who are aware of the number of these angels and the diversity of their assets through the historical indication of day and night. Without the permission of Allah, these angels cannot serve as intercessors in heaven or on earth.

What exactly is the mission of the intercessors? Where the forces of man and objects are inadequate and therefore the negative side of the balance increases, the intercessors intervene to keep it balanced. Allah creates everything for the benefit of people. Through the very presence of objects, how faithful people are in the state of their good-evil mobility, and how much they serve the truth is revealed. When the number of intercessors rises, an increase in interlocking geometric modules also occurs. Only in this way, they spread in space both over a wider area and thinner-thicker and color-tone differences.

Allah expects His servants to serve Him so that He may act justly towards those who believe. The most important thing they should do is to think. Thinking of how the sun sends its rays to the Earth and how it sends light, how the number of years can be known without the ranges of the moon... Because everything is created according to the ninth heaven, everything is real, that is, nothing is even an axiom. Those who refrain from Allah know that there is evidence of the change, lengthening and shortening of the day and night, which is depicted through objects. While the heavens and the earth were aligning, Allah separated them. One cannot have faith unless he sees these things and contemplates upon them. Therefore, the person's desire for knowledge and research is a precondition for any awareness to occur.

It becomes possible to prove that the earth and the sky were separated by knowledge. This means that everything has an explanation, answer, and solution. Allah offers us all the knowledge we need for our lives to be able to live in the universe. The West names this present knowledge in the universe as *"determinism."* Thus, seeing and touching knowledge is only possible through perception. To perceive something, one must carry all of the features of what he perceives. Another example that resembles this issue in the Surah Al-Anbiya' is that the sky is a protected ceiling. However, people turn away from verses of the sky. In fact, there is information about how the sky is able to be like this.

3 HEADING TOWARDS QUANTUM MECHANICS

3.1 ALLAH'S VIEW

3.1.1 HEADING TOWARDS THE SIDRAT AL-MUNTAHA THROUGH THE FATIHA

The Sidrat al-Muntaha means the last cedar or the cedar in its solitude. Gabriel (PBUH) left Muhammad at the Sidrat al-Muntaha and said, *If I go forward any more, I'll burn.* His characteristics were not suitable to be there because it was the last boundary where Gabriel could go. Beyond being excluded from any deficiencies, the Prophet Muhammad, who carried the strength of the knowledge of the names and attributes of Allah in the profoundest of ways, had an atomic convergence with Allah as much as possible. Allah excluded His blessed companion from the sins and weaknesses at the Sidrat al-Muntaha. In a way, this meant that Allah saw His own reflection in the direction of Fatiha where the servant was present.

The highest place a human being can reach is the Sidrat al-Muntaha. However, as the human servant remembers where he previously was in the spiritual realm, he advances towards that direction. To this end, one must first be in the blessed places of this world at heart and rise thus to his heritage. As the servant rises the seven levels, each step towards Allah in the servant's Fatiha increases, and the traces of the Sidrat al-Muntaha, which is quite far from the servant with each step, become clearer.

In general terms, the more the servants advanced in their own Fatihas within the single Fatiha the more angels they left behind numerically. This is because just as the Prophet Mohammed had left Gabriel behind, the same was true for all people or Muslims, since there are varying degrees of people and angels in regards to their faith (though their essences are the same). These varying degrees also take place between each other. Allah says that man has no knowledge regarding the creation of those angels. They have two, four, or six wings, and so on. And the servant stands next to the very angel he shares some characteristics with and they face the same direction. That angel, thus, becomes his guide on the way to Allah and is the protector of the human servant.

The universe and its atmosphere were created according to the form and style of people. The angels only appear temporarily in the world. They are always with us even if they do not appear more. If the angels had come out of the invisible world, the earthly works would have been ruined. *They say: "Why is not an angel sent down to him?" If we did send down an angel, the matter would be settled at once, and no respite would be granted them. (Al-An'am 6:8)*

3.1.2 EVERYTHING, DESCENDING AND ASCENDING ↑ ↓

He it is Who created the heavens and the earth in six Days, then established Himself on the Throne. He knows what enters within the earth and what comes forth out of it, what comes down from heaven and what mounts up to it. And He is with you wheresoever ye may be. And Allah sees well all that you do. (Al-Hadid 57:4)

*You see how you cannot
catch the moment
that time offers
in due time
on the mountain of Purgatory*

*in this duty of yours
you forget
heaven and hell
besides the veil that has been removed
from the middle of the mountain*

*kneeling, will you agree
to take
what you are to take
or is the past*

fading away
for this very reason

Everything is in the descending and ascending ↑ ↓ state. Since the Owner of All Sovereignty belongs to Allah, all things return to Him. The Owner of All Sovereignty is the sole owner of the universe. Allah's creation of the universe began with a single order *Be!*, namely *the Big Bang* in physics, and all the structural-systematic dimensions of this command were contained in this initial one. When we think of the universe as a garden, Allah determined what, how and at which dimensions and numbers things would exist, and He said *Be!*. Then, this command gave birth to other *Be!*'s.

Be! is Allah's expression, and it is a word. It is the representative particles and/or flows created in a way to respond to the word *Be!* accordingly. These representative units bow to this word and act according to its commands, so that their subservient capacities allow them to transform into each other. Formations and words arising therefrom reach to Allah. The essence of the formations in the principles of verbalization enables the relativity, which occurs according to the data potential in their modules, to be separated from the essence and to form a separate sphere of knowledge like the scales of justice. That is to say, as the statements correspond to the word *Be!*, the particles rise to the seven levels of the heavens, and the dualism operates reversely (a reverse perspective) and the truth gets separated from the superstition.

If this proportion is balanced in the entire human life, then the symbol of the mountain of Purgatory in between heaven and hell is revealed finally. The adventure of finding the truth and the faux isn't over or done. After they both reach Allah, He adds an abundance and science from Himself to the truth. They descend to the servant within the dualism, which will endure that dimension. Otherwise, there would have been no increase in the potential of faith or *the open path* of the human being. Allah is aware of everything, and therefore the reverse process naturally includes the ones below the seven layers of the earth.

In summary, the purpose of the reverse process is to reach a conclusion, or rather, many conclusions. The realization of these conclusions depends on the stage of knowledge scattered around. It therefore means that parts of information that are closely related to each other determined as the results have transformed into singularity. And the main purpose of all the singularities is to make statements return to Allah as a single word, delivering the building blocks of the hereafter to Allah for its construction and distributing the other consequences to the universe once again. It is the iron that enables the dissemination of knowledge as a result of the conclusions of these words to the universe. *The glance of Allah is one word.*

3.1.3 THE GLANCE OF ALLAH IS ONE WORD

The infinite harmony of knowledge is proportional to the distance between Allah and His servant. This diversity is circular-geometrical from the smallest unit to a single one, and it is integrated. On the other hand, because Allah is omnipresent -Ayat al-Kursi- His single view is reserved in all verses, in the unification of a point or many points according to the dimension of knowledge. When they touch knowledge, the particles from the glance wait to become apparent. Whenever the electrons in humans are under the influence of the sense of this glance (neutron), traces or its projections, the mind-heart will be fed according to it. All inventions of people originate from the presence of this glance in everything and in itself, because it is dispersed throughout the universe and it has layers. These layers are set according to a certain time period and thus combine human knowledge accordingly. In this sense, all knowledge brought by it carries the truth.

3.1.4 THE GLANCE OF ALLAH

The glance of Allah is His command, which resides within man and the universe. All things obey this command. His word shows how they should flow in their orbit. Whenever Allah takes His word out of things, everything ends up being destroyed. How has Allah created the earth without an axis? The sun and the moon obey Allah's command, as described in the Surah Ya Seen. As in all things, Allah breathes His own soul into humans who are the very microcosm to the extent required by the stages of man's creation. Adam, the first of all men, was first created from mud, and then, from unstable water, and at the third stage, Allah breathed His soul into him. Therefore, Allah gave His names and attributes to human beings according to their physical layers (multidimensionality), as well as the organ structures in their bodies and their functionalities. The main lines of these endless blessings are of course the knowledge of faith, mercy and love for Allah. This corresponds to the principle of existence. How can human beings transform something into feelings that is not flowing through their veins? How can they fall in love not even knowing the name of the beloved? In this sense, if Allah's soul was not in the body of man, there could be no harmony in the body and in the universe, namely, symmetry in terms of physics.

3.1.5 ALLAH WILL COMPLETE HIS DIVINE LIGHT

And [recall, O People of the Scripture], when Allah took the covenant of the prophets, [saying], "Whatever I give you of the Scripture and wisdom and then there comes to you a messenger confirming what is with you, you [must] believe in him and support him." [Allah] said, *"Have you acknowledged and taken upon that My commitment?"* They said, *"We have acknowledged it."* He said, *"Then bear witness, and I am with you among the witnesses." (Ali Imran 3:81)*

Allah is the only heir and He will complete His divine light. Completion of His divine light means to unify and complete the various units, of which all known and unknown prophets of all time were given the mission to serve upon Allah. The prophets fulfilled their duties. Since Allah has the titles the First and the Last, people must live in the aftermath of the last prophet and should be instrumental in fulfilling Allah's light by actualizing their narratives or keys, i.e., realizing their Fatihas, just like any other human being. In short, the completion of His light is the very point where knowledge peaks. In the hereafter, Allah will impeach even the prophets. In other words, the past religions—Judaism, Christianity and

Islam–complement each other as well as Islam itself. The same is also true for the prophets, all people and all things. They complement themselves and each other. This varies according to the situation, one's knowledge of Allah, and the perception of this knowledge.

If an ordinary servant, for example, sustains the characteristics of the Prophet Abraham, he will have those characteristics to the extent of their intensity. This is the case even for a single minute of the day. Thus, this person receives the very light that Allah has bestowed on the Prophet Abraham. Then, he leaves it again to the universe so that another one receives it and completes the history of this representative deliverance of faith. And yet, another dimension speaks for itself through the principle of integrity, which is Muslims are the ummah of all the prophets of the past, as they are that of the Prophet Muhammad who stood for unity.

Indeed this, your religion, is one religion, and I am your Lord, so worship Me. And [yet] they divided their affair among themselves, [but] all to Us will return. (Al-Anbiya' 21:92, 93)

The most basic word of Allah's light and that is in all things, even the atom, is the secret. Allah will cause those who try to extinguish His light to fail. So far, Allah has caused all attempts to deny Him to end up in failure. The best example of this is quantum physics. As is commonly known, the German religious and physicist Max Planck laid the foundations of quantum physics. Although he worked in many fields of physics, just as Albert Einstein had, he received the Nobel Prize in particular for his work in quantum physics. In fact, the Nobel Prize has been given to quantum physics rather than classical physics.

3.1.6 FLOWERS OF DIVINE LIGHT IN THE HEART

The exalted place of purity is the heart as Allah, who is merciful, can only fit into the heart. The point where people meet together is compassion. The Prophet Joseph recited the cycle of mercy in the Qur'an by saying: *I do not purify the nafs because the will orders the evil.* In a body where the desire is intense, the protection of people from evil is only possible by strictly following Allah and asking for forgiveness from Him.

From a different perspective: The servants must build their bricks for the Kaaba in their heart to be the voyagers on the right way. The Kaaba is constructed based on faith and Islam, is raised up and formulated accordingly. For example, if a servant sees Allah in his prayers, the time and place disappears, and one's heart sets off on his journey to Allah. Why the heart? This is because the merciful Allah can only fit into the heart. The existance of compassion in human beings may take place but only by turning to Him, and thus, imperfections are forgiven. But not by worshiping the devil... In that case, the servant builds his life upon evil by entering into the service of Satan. Running after idols and ending up harming oneself...

The merciful Allah has poured His wrath on the rebellious Satan. He also punishes those who follow Satan and causes them to forget Him. Man should refrain from being within a community that insists on disbelief in Allah. The reason is that Allah delays granting some favors to those servants who are in such an environment. On the other hand, those who do not have such an awareness of faith, such as Qarun and Pharaoh, deem success as an outcome of their own intelligence, so Allah immerses them in grace to increase their insatiateness. They ignore the truth, rather than seeing where and whom the grace comes from. They boast throughout the earth and worship anything but Allah. Before Qarun, who became rich for his servitude to Pharaoh, imposed disbelief upon his society, the Prophet Joseph had achieved tremendous success by Allah (approximately 1,800 years before the Christian era, when there was an extreme worship of idols in Atenism).

3.1.7 THE PROPHET JOSEPH AND THE DIVINE LIGHT OF ALLAH

The sun rose upon my face
oh, my favorites
what of you is left for me

those days I was there for you
now I am all alone
with what I couldn't accomplish

Just like the way I once changed

I became something else and
grinding stone

I have left what I lost in surrender
and to past seasons
the alms of the soul never ending

if they had given a fortune to a greeting
it would have lost nothing from such a secret
without dying while still serving

Some of the surahs of the Holy Qur'an begin with cognate verses such as *Lam, Ra, Mim, Sad*. These words, which are presented to the people so that they can know Allah, are closed to interpretation according to the verses and He holds their judgment and secret. However, change and variability are essential within Allah's system. For this reason, when the time of the cognate verses comes, they become steadfast. Those who learn from this wisdom are wise men that have reached a high state of knowledge. Allah bestows his own secret upon them. In other words, the cognate verses become steadfast verses by the will of Allah.

As an example, we are able to say that while in the past only classical physics constituted steadfast verses, today they have first become cognate verses, which have moved towards these universal dimensions, as quantum physics has become steadfast now. The beginning of surahs in this way concerns each of the verses in a singular and holistic principle. The universe is created in harmony by the characteristics of each person in the system of knowledge. Therefore, along with the hereafter, the universe appears to the person as he knows it. In this sense, everything starts with the known and the unknown. However, the common feature of these both is that they are both verses. The first one is visible and the other one is invisible. When a person finds himself in a situation, fear arises because he does not fully understand the degree of his own ignorance. Just like in the story of the Prophet Joseph... When the siblings of the Prophet Joseph threw him in the well, the prophet became afraid and there he received knowledge, i.e., a revelation, so as encourage the prophet in response to his fears. This is an indication from Allah ensuring that he is safe, he will overcome his brothers, and therefore he should pray.

Egypt is the country where Allah appointed the prophet Joseph for his duties. Before performing them, the Prophet Joseph needed to learn and understand the system in Egypt so that he could fulfill his duty appropriately to Allah's command. The status of the voyagers on the right path arose even while they were still on the earth. The reward of beauty will never be lost in both worlds. Everything is but for a reason.

Through the command of Allah, the Prophet Joseph was separated from his father at a very young age. Although his father, the Prophet Jacob, learned about his son through revelation, he said that *he had told Allah only about his grief.*

On one hand, there was knowledge and prophecy, and on the other hand, there was sadness. We can clearly see that the basis of this contradictory behavior lied with the fact that the prophet was a human being. In short, there was a commitment to dual valiancy; a prophet doesn't have a lifetime or even a certain period of time to remain pure. The Prophet Jacob is a good example of the idea of being created through this dual characteristic through the story of Joseph. When his other sons took Joseph away he told them he was afraid that a wolf would eat him. Yet he knew that Allah would protect his son. The Prophet Jacob became blind because of his grief and sorrow due to being separated from his son, Joseph. However, he knew that if his other sons thoroughly should look for him, they would find Joseph in Egypt, and adviced them to set off. He always reminded his children of another attribute to be learned, the expectance of hope from Allah, which is never to give up hope. Otherwise, the hope for His mercy would have been lost.

3.1.8 CHILDREN, SCIENCE, KNOWLEDGE AND THE SURAH ASH-SHARH

People want to have some things in their lives. Sometimes these are worldly things like houses and cars. Sometimes there are desirable situations for both the world and the hereafter, just as living and dying for the sake of Allah, like in the *rabbana atina* prayer. The prophets, with their messenger and nabi identities, are human beings and therefore, they may have worldly desires. Allah, in some verses, even warned Muhammad Mustafa to avoid the temptation of envying them. In addition to insignificant requests like this, the desires of the prophets are in accordance with the *rabbana atina*. This is because: *The example of those who spend their wealth in the way of Allah is like a seed [of grain] which grows seven spikes; in each spike is a hundred grains. And Allah multiplies [His reward] for whom He wills. And Allah is all-Encompassing and Knowing. (Al-Baqara 2:261)*

What is spent on the path of Allah? It is the zakat and alms given on this path. It will be wrong to see alms only as financial aid. Alms-zakat are to realize the blessings of Allah and the way of being thankful. Just as the blessings of Allah are endless in abundance, there is so much variety of alms. At the very foundation, the servant must direct his life, walk on the right path according to Allah's names and attributes, and fulfill these mighty attributes as much as he is able to. Thus, he gives enough alms to Allah. His servitude to Allah and alms-zakat are to worship; to love himself, his wife and child; to serve the path of Allah; to educate individuals who are beneficial to society; and to work faithfully. And there are seven hundred kinds of beauty in response to one beauty... This one beauty carries within it seven different faces of beauty, and then it becomes a basis for other beauties.

Considering the desires in line with the *rabbana atina* once again, having a virtuous son who serves Allah is among the greatest desires of the prophets. Allah responds to these wishes when the prophets leave their environments, those environments that are unable to be restored and predominantly made up of infidels. A good child can be born in an environment where evil no longer exists. A good child gives hopes in intercession for humanity, the mother, the father and the relative. It is a characteristic of servitude, which Allah loves and can be given as the intercessor's right to speak on the Day of Judgment. He is the one who can be an intercessor for parents, relatives, even friends and angels and other things. This is because the angels revolve around this servant and ask for mercy in the presence of Allah if the servant has any flaws. And the angels and things are also pleased with the servant who loves Allah.

3.1.9 MOTION

We thought that nothing happened. However, is this ever possible? Everything has been created to be and remain in motion. The state of creation is the source of all kinds of creations, even if it is seven times as deep as a sea of ink, and if it ever were to dry up, Allah's creation would not end in the Fatiha of the seven worlds. In the case of being created, we know only the part of the future where we are returning to and stuck in the repetitions of our preferences. And those are the cycles of resignation and acquisition formed from thin and thick shapes acquired from the values we give them and the life we are dependent upon. These cycles are always voluntary as long as we are breathing. Things appear to be absent or involuntary. The reason for this is that these parts have remained far behind because only particles of these parts have been touched.

In other words, since we have touched the objects and their dimensions, which are a part of us, the differences we find or notice in the people around us are the results of us touching these characteristics less. After that, our reaction to these differences comes from an effort to alienate. With the proportion of the discomfort felt by such a thing and as that thing becomes more and more apparent in the person himself, it becomes more evident to him. The attitudes and behaviors towards the phenomenon of such a thing, which people feel strange about, becomes more sensitive. From here, knowledge is born or it becomes practical for phenomenon to be information. Since we are all a part of creation, and everything and everyone created is within us, then we should not be judging our preferences. Not judging such things is the proportional patience that responds to each person's attitude and behavior. Proportional patience means the adjustment of patience and a reasonable temperament based on the attitudes and behaviors encountered. From here, it also means to respond to the entirety of the attitudes and behaviors of others. Only in such attitudes and behaviors, i.e., expressionlessness, the original expression is visible.

On the other hand, in the movement and mobility of the universe and human beings, there is a requirement to touch all objects, whether they come to mind or not. Without this, the creation of man would not have been symmetrical. More precisely, the symmetrical understanding of the human is not necessary. The more awareness we have, in other words, our ability to read the very manifestation of the names and titles of the exalted Lord of the universe, the more superstitions will disappear, and the only successor of the earth, Allah, will appear.

At this point, the idea that Allah has made all Muslims His witnesses and that He can only fit into the human heart shows that He is the only heir of His servants. What is more, because of his unique generosity, Allah also bestows His attributes upon men. *And We have already written in the book [of Psalms] after the [previous] mention that the land [of Paradise] is inherited by My righteous servants. (Al-Anbiya' 21:105)*

3.1.10 ALLAH HAS GIVEN YOU THE NAME 'MUSLIMS'

The emergence period of Islam was a revival against the period of ignorance and its systems. Adam's creation was complete. This means that Allah included all the names of the objects into Adam's creation. That is, while believing that the first thing to be created was the light of Prophet Muhammad, the first human being and the Prophet Adam was the very deductive expression itself.

The creation of man begins to take shape with the saying of *Kûn (Be!)*. The completion of this beginning is the word of Allah, and the verse of creation in the Qur'an says that people's ambitions are written in the book hanging on their necks. Allah knows how He created His servant. He wants to declare this and inform the angels of his power. In front of the congregation, the angles asked the Prophet Adam for the names of all things. The Prophet Adam said the names of all things. Allah asked the names, i.e., He wanted them, and the Prophet Adam knew and answered Him by the very desire (want) of Allah. Here, it becomes obvious that the meaning of asking is the answer to the questions.

Every period and periodicity has its excesses. If the period of the Pharaoh and Qarun were an invitation to the resurrection, this would take place through the Prophet Moses. The reason for the coming down of the books and the arrival of the prophets was to bring righteousness to the negative conditions that were in power, i.e., superstitions. For this reason, the prophets revealed the intricacies that created awareness in human beings through the means of struggle and announcement.

The common code of all periods resulted in the intersection of worship and being worshiped. The Qur'an's longest surah, the Surah Al-Baqara, dealt with the behavior and attitudes such as arrogance, which greatly affects a person's life and enables the growth of things devoid of truth. From an earthly point of view, the first man and prophet is Adam. The last prophet is Muhammad Mustafa, though not the last human.

Adam represented all of the people *(Our creation and your resurrection will not be but as that of a single soul. Indeed, Allah is Hearing and Seeing.)* *(Luqman 31:28)*, but the Prophet Muhammad Mustafa (PBUH), who lead people to the truth, who was the best way to know the uniqueness that can occur in human beings, and who will take man into this uniqueness, was a gift to mankind. It is useful to refer to two specialties here. The Prophet Moses and Allah spoke to each other. In other words, Allah himself spoke with the Prophet Moses on Mount Sinai, and in another story, He fulfilled Moses' desire to see Him. Allah appeared on the mountain so that the heart of the Prophet Moses could most intensely feel love for Him. Seeing this, the heart of the Prophet Moses could not endure this love and he fainted. The heart, especially the heart of a prophet, is stronger than a mountain. Thus, the mountain shattered as Allah manifested. The might of Muhammad Mustafa is that he ascended to miraj while his eye was on the Sidrat al-Muntaha without crossing the border. He could see even in this world what no one else had seen or could see in the hereafter. This is only possible with the holy Qur'an that Muhammad Mustafa (PBUH) announced to mankind through obedience and the worship to the Heir of the Heirs, the Most Gracious and the Most Merciful Allah, through representing the will of Allah, and surrendering himself to Allah. Muhammad Mustafa (PBUH) became the Fatiha. The Qur'an's verses, words, and all its truth unites with the formation and particles of the road and becomes one, and thus the last prophet ascended to the sky from the Masjid al-Aqsa and the Masjid al-Haram expressing the light of the earth. The superstitions disappeared and his surroundings were lightened with the divine light of the prophet like *foam that occurs on the water and flows away.*

3.1.11 HUMANS ARE A PART OF EVERYTHING

Say: "If the whole of mankind and Jinns were to gather together to produce the like of this Qur'an, they could not produce the like thereof, even if they backed up each other with help and support." (Al-Isra' 17:88)

This verse emphasizes the place and characteristics of humans in the universe and underlines that human beings generally cannot kill anything that is not in the universe. In other words, people can only perceive the tree as a tree with the qualities of a tree and an apple tree as an apple tree. A number of objects thus exist (we are also an object, so other objects bring us into being by perceiving us). Humans do not have anything that is not there in the universe. Everything that happens in the universe can be found in human existence.

Let's take a look at human chemistry here. There is no lacking in the chemistry of the human who was created from the dirt. For instance, iron, copper, etc. should all be present in sufficient amounts. It is only thus possible to obtain the necessary biological interactions. In a religious context, humans are a piece of all religions, such as Islam and Christianity. From a historical point of view, Allah carried the whole history of mankind, as he is the First and the Ever Enduring, and He breathes His soul into others.

History encompasses all people who have ever lived and existed in the universe. Therefore, we get to know their existence in us through the knowledge we have gained through the Prophet Muhammad as well as from all people in the past.

On this subject, in his book entitled *Wholeness and the Implicate Order*, published in 1980, the physicist David Joseph Bohm stated that everything in the universe must be understood as part of an integrity that is undivided and in flowing motion. Bohm's definition of universal functionality was a frequent common expression used centuries ago in the Sun, the Moon and other objects in the Qur'an. The act of flowing and swimming already includes the meanings of existence and the absence of a structure in motion. Punctum and momentum can exist just like that. In other words, the acquisitions of the truth gained through experiencing the existence of the point within this dichotomy, according to the proportions of the situation, correspond to their dimensions and dimensional relationships.

3.1.12 ROLAND BARTHES: PUNCTUM AND DETAILS

And We have certainly created for Hell many of the jinn and mankind. They have hearts with which they do not understand, they have eyes with which they do not see, and they have ears with which they do not hear. Those are like livestock; rather, they are more astray. It is they who are the heedless. (Al-A'raf 7:179)

In the verse above, Allah says that He did not create sense organs in vain and each sense organ has a specific function, but the infidels violate these duties by not using their sense organs. The importance of sense organs is undeniable. This is because, if there is a heart in the middle, this means there is a large volume of space where the sensations of the intense compounds of the senses are combined to form the most intense emotions. For example, through the organs such as the eye and the ear, the universe is seen and heard. Thus, when the veils, which represent artificiality, are removed, Allah becomes visible; because the heart can only be understood through the help of these sense organs. They contain the truth. They have the ability to understand everything that has been created. And as a representative body of that truth, they have the potential to terminate all kinds of effects from internal and external factors as well as the tension caused by unknown energy adapting to newly established balances.

The conclusion we arrive at here is that the weightier the unknown energy is the more detail is left impressed on the heart. Let us elucidate this concept of detail by looking at it from a different perspective. The French philosopher and semiotician Roland Barthes clarified his ideas regarding photography through the terms punctum and studium in his book *Cameralucida*. The brief meaning of the term studium refers to the effort the photographer puts to make sense of and evaluate cultural-social knowledge. Studium is an interpretation of a particular situation from a biased point of view. According to Barthes, when we look at a photograph, as he puts it, the thing that *leaves an impression on us* is a particular detail that affects us for a moment, i.e., awareness. It is not possible to define this detail, but it becomes meaningful in our heart, which presents it to us so that we can perceive it through our sense organs. The universe is a photograph adorned with endless photos. The fact that a person is always involved in detail means that a photograph appears out of another one.

In fact, since everything is chemically related to creation, everything that is familiar to us is punctum. This concept refers to a promise since it is completely about Allah's single look or word. The detail we take from there is the source of our moments, i.e., our truth. And other parts in the picture have to take place in the background. This is because the detail covers the picture, so meaningfulness and meaninglessness occurs, that is, the space-time unravels. Thus, punctum is not only a means of living out truth, but also influences it. In this context, punctum is the taste of divine love with its sweetness and bitterness.

3.1.13 TRUTH

Truth carries laws relentlessly. While carrying them, it establishes the balance of the universe. It protects itself along with the balance and harmony. From then on, the laws take the form of what we believe and guide humanity. If the laws were to carry the truth, it would be natural, but this is not the case. This is because laws present situations. They are also dependent on these situations, and they rewuire decisions to place them within themselves. However, the particles of truth should be distinct, or at least sufficient, within the decisions for these situations, so that they (in the form of light, matter or a word) shall not be sent alone. Moreover, the truth should turn again into laws with necessary additions, thus making systematic artificial laws deficient in governed places at least to some extent.

3.1.14 THE *MATHEMATICAL* RELATION BETWEEN WORDS

Let's take another example: Does the egg come from the chicken or the chicken from the egg? We can see that this is not much different from what is known to the public and has become thought material regarding the material-spiritual structure and the creation of the universe for centuries. Now, the question is: Was the atom derived from the word, or was the word derived from the atom? In this case, would the chicken-egg example be turned upside down? Each question has its own answers that determine their outcome. In other words, solutions are relative (= time), but the energy that consists in the overlapping and intersection of space-time and the intensity and depth of the geometric shapes (without the friction) from the geometric shapes mentioned, give light to the universe and beyond.

This light is pure, innocent, the very truth itself. The remaining space-time continuum uses this light. It is the source of different good-evil conditions, as well as offering particles to truth's differences. It gives power to the truth from truth's own power. As it is understood, forgiveness of good-evil (dualism), which are the atoms that make up the universe, is required to return the existence or reflection of the truth it carries within the space-time continuum to others. The aim here is to extend the space-time veil as much as possible. The removal of the veil means that the particles of truth will appear in the space-time continuum as they touch good and evil through their power. However, with the overlapping of the merits of truth (knowledge, love, respect, justice, compassion, innocence) along with reality, it continues to go on its own way in the universe, and thus expands and deepens its own path. Hence, both the chicken comes from the egg and the egg comes from the chicken.

Such examples *as It is easy for Allah* are given in the Qur'an: The Prophet Adam, who was created out of nothing; the Prophet Jesus, who was born without a father; and every other person who was born with a mother and father... Hence, the only ruler here is Allah, and it is man who is obliged to reach all kinds and forms of awareness through repetitions. The Prophet Muhammad, representing singularity, came upon this legacy by fulfilling his duty of prophecy and ascended to the miraj by unifying the truth among truths and he had also been at the Sidrat al-Muntaha. On the other hand, various stages of the Prophet Muhammad (PBUH), who was sent to people as the last prophet and grace, and other prophets and believers are delivered to him through revelation and these reprisals are repeated in the verses of the Qur'an. These kinds of stages refer to repeating what they have known from time immemorial and delivering the same love in an invisible form. The last prophetic period is the culmination of goodness and evil, which are also reflected in the universe. The last point reached is ignorance and it is emphasized that ignorance is actually the nest of evil. Muhammad Mustafa (PBUH) was a gift for all people. His Ummah that are walking in his path resemble the prophet and become him to the extent that they follow his path.

The goal of the servant is take the faith of the believers as an example and to live it out and to reach Allah, who is not subject to deficiencies whether in time or space. In order for this, the common path and purpose of them all is the straight path, i.e., the Fatiha. The prophet knew the Pharaoh and Qarun. The Prophet Moses and the Prophet Aaron were informed through revelation to oppose this evil and its systems, which were created by this evil. They took shelter in Allah.

The Prophet Muhammad knew the Prophet Solomon's perfect knowledge of repentance. And to this extent, he knew the unlimited abundance Allah can bestow upon him and the believers. The Prophet Muhammad knew about the Prophet Abraham's battle against idols. He knew that mankind could only avoid illiteracy through worship, and he fought from the depth of his heart for the Muslims, who learned about and recognized the unity of Allah, so that they would multiply. Allah begins the first verse in the Qur'an, with *Read!* in opposition to illiteracy. *Read in the name of Allah, the Most Gracious, the Most Merciful.* This is because everything is created by His word; everything is a verse, i.e., a decree.

Reading in the name of Allah means to *Believe, believe so that you can see the truth, and look around, that faith will be strengthened. Look around again, see if you can notice a grace. You can see the impeccable creation of Allah, who has built the sky above the earth without a pole.* Abraham expressed this in the most profound way. In one of the most intense periods of idol worship, Abraham set off to find Allah with his faith blossoming inside. He realized that Allah was not the greatest star, the Moon or the Sun. He realized that there was a single Allah who created everything out of nothing.

3.2 WOULD THERE HAVE BEEN *BE!* WITHOUT ANY VERSES?

3.2.1 HUMANS AND THE EARTH

The fact that man was created from the soil is one of the proofs of Allah's existence. *And of His signs is that He created you from dust; then, suddenly you became human beings dispersing [throughout the earth]. (Ar-Rum 30:20)* To achieve equality here, gravity according to the type of soil and essence of humans is created at a density to adapt to this gravity. The meaning of living by the principle of equality is for a person to determine which land he will live in, i.e., his fate, himself. This is biologically relevant to all organic and inorganic beings. The chemistry of the parts is to answer each other's biological expectations and complete their deficiencies. They give signals to space and time in search of incomplete knowledge in themselves to ensure their integrity. With the power of gravity, whether close or distant, space and time respond to these requests in the direction of the asymmetric and symmetric calculations of mathematics.

3.2.2 CREATION AND MERCY

Allah has created objects, animals and people according to their own special measurements. To be created based on these special measurements is to realize the shape of the action and reaction of the Fatiha given to this measurement through its chemistry. The Qur'an describes this subject in the Surah Ya Seen through the Sun and the Moon, which are moving in their designated orbits. Since everything and everyone created has a trajectory, the soil also travels in its orbit. Allah has given the soil the authority to call the human to account, who was created from it, then the soil has the consciousness of such account it carries within itself. This consciousness is the Fatiha, which is suitable for the characteristics of the soil. The conclusion is that only the soil, which contains the names, titles and the Fatiha of Allah in its own chemistry, has the authority to call to account; if it has the authority, then it can also do a task. Allah created all things and enabled them to create cycles. As a result of these cycles, the creatures themselves take on different geometrical shapes such as triangles and squares, and give shape to everything else that creates life, i.e., all things that are involved in their creation. In any case, there is an interaction so that all objects know each other and the things they are doing. This has to be so, in order that every requirement of the balance should return with the pluses and minuses.

This situation may arise from the simultaneous realization of three different situations: Matter rotates on its axis; as it rotates, it also rotates around everything; and as it rotates around everything, so does everything in the universe. Here, there should be an essence that needs to ensure this configuration and it has to bear the characteristics of the shapes it will make, just like the atom and the light that it leaves on all the objects in the universe.

Allah has presupposed to add one thing to the Fatiha that exists in the atom, to all His mighty names and titles, and even to Himself, which is mercy.

There is always an accusation that all the prophets face. The objections raised by hypocrites and disbelievers at the beginning are that the prophets are also human beings just like them. However, the faith of man, who is in need of Allah, becomes meaningful through the fate of being a human. Therefore, in the destiny of a human being fed by delusion and arrogance, there are times where it leaps out of its orbit of Fatiha. These are inevitable because people are human beings. One performs his rotations as an interchange, but in a way that his hands are tied. According to the velocity and strength of the leaps, it sprinkles particles around itself and to the universe from the Fatiha.

They have to be scattered for only then will it be possible to combine the Fatiha from its particles through knowledge. Scientifically, this is called the connotation pathways. In other words, a person who is created with an asymmetric structure creates his own mobility as much as he lives depending on space and time. And this mobility creates orbits for the space-time continuum so as not to lose its movement to the extent required by the ebb and flow of good-evil dualism and the space-time continuum. In other words, it makes its own orbits. This shows that the functionality of the human or the universe is perfect, symmetrical. As a result, due to these orbits, the force of gravity provides a balance of living in man, as needed. The gravitational force simultaneously leads to a reduction of things from people, while still maintaining the balance. (Qaf 50:4, 5) This is because the force of gravity provides the balance of vitality in humans. It collects the necessary chemistry and ingredients for this balance, leaving the rest proportionally close to the human being in a distant time and place. With these deficiencies, chemicals that cause aging, sickness and dying cannot change their trajectories.

3.2.3 MATTER AND SHADOWS

Have you not considered your Lord - how He extends the shadow, and if He willed, He could have made it stationary? Then We made the sun for it an indication. Then We hold it in hand for a brief grasp. (Al-Furqan 25:45, 46)

Allah asked Muhammad if he had ever seen how He shortened the shadow, i.e., whether man learned a lesson from the shadow. And in line with the extension-drawing principle of the shadow, Allah mentions two important things: First, the Qur'an fulfills all the needs of quantum physics. Let's give a simple example. The entire object has a shadow and these shadows turn to the left and the right. This moving state helps to change their dimensions. The size of the shadows change depending on the object and the environment they are in. They also allow the item to change according to varying perspectives. This illuminates Einstein's theory of general relativity.

Another reason why shadows have varying dimensions lies in their ability to allow transition between day and night. More precisely, they greatly contribute to the existence of natural flow, i.e., the seasons, the wind and the rain. This is because shadows play a decisive role in where the spectrum will fall, and with the extension principle, the spectrum particles get dissociated and distributed to the dimension where the shadow is extended.

He Who created the heavens and the earth and all that is between, in six days, then He established Himself on the Throne: Allah Most Gracious: ask thou, then, about Him of any acquainted (with such things). (Al-Furqan 25:59)

Allah's title of the Most Gracious is mentioned in this surah. There are those who know this power, and of course, they know its might, its greatness and its knowledge. In other words, in the verses of the Qur'an with different explanations, the servant reads about and learns how the water was taken up to the ninth heaven. Not only ascendance of the water to the ninth heaven is explained differently, but also that dimension of knowledge, being included in the verses that explain different events, becomes constant in the consciousness of the person. Thus, one has a certain potential to provide his own movement (long-short) period from his numerator and denominator knowledge.

For this, there are conditions to know the Most Gracious. One of them is, of course, prostration. From another perspective, Allah, who presupposes mercy to Himself, has placed the nobility of mercy in all His names and titles. The basic function of good-evil in the servant is to ensure that the faith, which settles in the heart of man, becomes clear, as well as the more he touches the space-time continuum with different dimensions, the more he returns the particles of truth (Fatiha) to Allah who owns the truth. (The heirs that Allah mentions in the Holy Qur'an.)

Every particle has mercy in it. Mercy is to love the creature for the Creator's sake. Mercy is that the neck is thinner than a hair. It is the hope of Allah's forgiveness. Even the angels beg Allah to forgive His servants. The man who is in need of mercy knows no penitence if he is merciless. In this case, the particles of the Fatiha cannot be combined together. They stray away, and thus, leave no place for the Surah Al-Maa'oon and the Surah Al-Fatiha to exist; they cannot complete each other in this respect. This is because the person who is distance from mercy cannot be grateful or repent. Without them, the need for alms and charity and their perception would be lost. However, material-spiritual zakat and alms prevent the repetition of defective attitudes and behaviors through purification and sublimation.

Allah, the owner of the ninth heaven, raises degrees. Do the degrees increase with the ninth heaven? The revelation about man's will descends to cause fear in them for the Day of Judgment.

Allah sends down revelation to the heart of the servant He wants. This will is the will of Allah in His names and titles. A remarkable situation regarding sin in the Qur'an is that Allah first forgives sin and then accepts the very repentance He has offered. That is to say, Allah grants His servant forgiveness after repentance. Repentance is the greatest grace given to the servant. Blessings are given to the servant who repents through grace. All kinds of blessings come down from the sky to the servant as rain, sun, stars, iron and so on.

3.2.4 THE SURAH AL-MAA'OON AND THE FATIHA

The Holy Qur'an is the Fatiha and it is in the Fatiha. What about the Surah Al-Maa'oon? The meaning of *Maa'oon* is to give something to someone as alms or for temporary use. In other words, here the importance of help is emphasized.

The Surah Al-Maa'oon is composed of seven verses just like Fatiha. The things that a servant must do to earn the love of Allah, but still doesn't, are summarized in order in the Surah Al-Maa'oon and Fatiha. Of course, the first verse in the surah begins with a question directed at Muhammad, Muslims and all people: Have you seen those who deny the religion? Who denies this religion? Who doesn't believe in Allah? When are they considered to be faithless? They have been described as those who ostracize orphans and do not encourage feeding the poor. And what is more, they do their prayers. But they do not know that the principle of prayer requires helping people and those in need through Allah's sake. Can their prayer be sincere within the framework of these attitudes and behaviors? No, they pray for show and they do no good, and they also prevent others from doing charity.

3.2.5 WHY OBEDIENCE AND FAITH?

Allah, in the exalted book, says *Give alms before you talk to Muhammad.* The Holy Qur'an is eternal, the last, everlasting and evident. It contains all the names of Allah. So everyone, that is, all the servants of Allah can speak to the Prophet at different times and in different places. But it is necessary to give alms before speaking with such a blessed person. Since the Prophet Muhammad is the most blessed, he is presented here as a representative example. In this context, believers may be able to negotiate their faith by opening the veil of space-time. Not only with the deceased ones, but also with the future blessed will it be possible to meet. The ones, who build their homes in the hereafter when they are on the earth, flow through time along with the holy people only by remembering them and following their beauty.

One possibility is to honor the surrender of a particle that is missing in their legacy. Therefore, the exalted book mentions the importance of faith, faith in Allah, the prophets and the angels. This is what Muhammad Mustafa followed in his circumcision, because those who follow in the footsteps of these beautiful worshipers will reach Allah. On the way to Allah, deviation is kept as limited as possible to survive.

3.2.6 THE CREATION OF LIFE WITHIN AN OBJECT

Man lives depending on space and time. This is the case for all of us. Situations, events and people are all one and they take us under their influence. As a result, they shape us. Our role is to play a role in how and to what extent this pattern will take place. As it is shaped, our structure insists that our choices are strengthened in the same way as their predecessors and continue the way of life they are not stranger to. Therefore, other people have other preferences. Then, other forms of life, attitudes and behaviors seem strange to us. They even become violent, and we find it funny and absurd. However, we also have the same things within us, but we have neglected these things. The only difference is this. When this occurs between countries, the marginalization increases. It is therefore essential to live in space and time, to understand, love and respect as much as possible, rather than drifting through space and time. Allah has created people's potentials and their every element in balance. And whatever the choice is, Allah is merciful; He is the one who responds to gratefulness.

3.2.7 THE SURAH AL-ALAQ AND KNOWLEDGE

Learning and comprehension: The child looks at the sun and asks, *Mother, what is this?* His mother answers, *This is the sun.* Then, the child asks everything he wonders, learning about the Sun. He understands that the sun creates warmth, rotates behind the clouds even when he doesn't see it. Therefore, as he touches the Sun's dimensions, the dimensions of the comprehension of the child also get more complex. In the end, much of the material has been included in the unique knowledge of the mind and the heart of the child, who is aware of the Sun's existence and mission, and thus is able to talk about the issue. For example, a child's stages of learning about the concept of a grocer are as follows: The first thing he learns about concerning this concept may be knowledge about what is sold in a grocery store. Therefore, the word grocery integrates with the words fruit and vegetables, and the child grasps that there must also be some lemons, which his mother needs.

In the child's first one or two visits to the grocery, his mother accompanies him. Meanwhile, the child also understands the concept of the grocery.

In his mind, both the mother's explanations and his own observations, as well as various pieces of the grocery and the grocery itself are now shaped. One day, his mother sends the child to buy lemons. The child either asks questions to make sure he has understood or he trusts his knowledge and does not ask his mother where he should go. If he asks, he will go shopping without having to ask questions over time; that is, if there is a need for apples or lettuce at home instead of lemons, the child will find solution in the grocery. One day, when he sees that there are no oranges in the grocery, he will find out that supermarkets also have fruit and vegetable sections. After all, he has been there with his parents before. And he has learned that some fruits can be obtained only seasonally. All sense organs are active in the child's understanding of the outside world, and these senses develop as the child learns. In other words, the outer and inner worlds grow together, and thus the existing habits and judgments in the child become evident. So, either the child wants something or he doesn't. Likewise, either he likes it or he doesn't. This dualism, which is the very vitamin of attitudes and behaviors, gives mobility to the child by being the source of both life and energy. This strengthens the child's commitment to time and space.

There are always connections with the child's experience and understanding, and this knowledge. Rather, the knowledge of the painter who fills his white board with his life becomes apparent only through touches within his mobility. Therefore, there is always knowledge circulating in his vessels. It is the one and only reason for life of the human being, who was created from unstable water. Man is as much eternal as his knowledge, because knowledge is a fact and its entirety is the glance of a single piece of knowledge. This knowledge holds all pieces of knowledge and greetings. These greetings and knowledge are the very information that returns to the object. Objects and angels do not get tired of chanting to Allah. They give themselves to Him. However, Allah created man. And when He told the angels that He would do this, the angels reacted, as they didn't understand it.

They knew that the so-called human being would shed a lot of blood on earth. On the other hand, Allah did what He said and created Adam. *And He taught Adam the names - all of them. Then He showed them to the angels and said, "Inform Me of the names of these, if you are truthful." They said, "Exalted are You; we have no knowledge except what You have taught us. Indeed, it is You who is the Knowing, the Wise. He said, "O Adam, inform them of their names." And when he had informed them of their names, He said, "Did I not tell you that I know the unseen [aspects] of the heavens and the earth? And I know what you reveal and what you have concealed." (Al-Baqara 2:31-33)*

For this reason, Allah knows who, when, where in the universe and to what extent lights up; he even knows about particles.

On the one hand, it seems that there are people created who oppress their own souls; on the other hand, the ability to learn Allah's powers through the knowledge of Allah and to not follow evil have also been given to these people. It is obvious that we cannot have perfect attitudes and behaviors. In fact, due to these deficiencies, mobility, which is the storehouse of knowledge required for consciousness and the subconscious, is kept in motion. But why does Allah want people to have symmetrical creation from their asymmetric structure? Does He only want to be known here? Or is it that Allah serves people so that His servants can refrain from rebellion and instead, reach the stage of contentment? Thus, those who are pleased with Allah attain to His consent, which is higher than all the heavens. What a generosity, what a love, what a greeting this is! For this reason, one can only keep and carry the love of Allah or knowledge of Him in his heart.

3.2.8 THERE IS ALWAYS SOMEONE WHO KNOWS BETTER

In the past, people once asked Allah to increase the distance between cities. Allah fulfilled their wishes, however, stating that such a demand actually brings nothing but harm to themselves. This is because distance means that people move away from each other spiritually. Cultural differences would occur depending on the region where they lived. Human diversity, the increase of scientific disciplines, the sub-units or sub-branches of the disciplines themselves show this. The rise of technology, the internalization of different traditions of religion, sects and regions within a country are some examples of this. At the same time, an increase in diversity means the multiplication of dualism. This means that accessing basic knowledge will require more effort at the end of life, so that the miracles of Allah can be clarified. Those who do not know this always want miracles, but those who know can see a miracle in the smallest thing. They are always determined and they search, examine, and read. Allah reveals His signs, and glorifies them.

An example of this is the fact that Allah tests the Prophet Abraham with a number of words. The Prophet Abraham is on his way to find Allah. He asks if the Sun and the Moon were Allah. But when they set, he says, *I don't like them setting,* and he exonerates Allah from deficient attributes. When someone tries to find Allah and wants to know Him, Allah will make Himself visible. If a person has acquired knowledge in the direction of faith and if he is still doing wrong, then he will give Allah an account for trampling on the truth. If Allah has given wisdom, it is necessary to show attitudes and behaviors according to this science.

3.2.9 VERSES, *BE!* AND ATOMS

Kûn means *Be!* Allah is constantly in the act of creation. After all, there is life in which the universe and the human breathe. When Allah says *Be!*, gravitational force, which is itself a balance, distributes balance by carrying all kinds of change in balance. The word *Be!* is knowledge. It is the knowledge of the name Allah, His names and attributes, and the intensity of His enlightenment. *Be!* is the order of how atoms should move according to the mathematics of an atom. It is understood from here that things and people revolve by Allah's will.

His will is the straight path. One must surrender himself to this will. There is no doubt that Allah shall fulfill His will to distinguish right from wrong, which is given through His authority to the person who is worthy of dignity. On the contrary, there is an increase that will lead to righteousness. Existence occurs in different dimensions of knowledge in the space-time continuum when Allah says, *Be!*. According to these differences, it taekes geometric shapes, its meaning becomes vast and deepens. *And [He created] the horses, mules and donkeys for you to ride and [as] adornment. And He creates that which you do not know. (an-Nahl 16:8)* From the lack of knowledge, the vehicles and their requirements are not yet available. This means that for the invention of that vehicle, the time assigned by Allah will fall into space on His timeline.

3.2.10 THE SCHOLARS, MUHYIDDIN IBN-I ARABI AND MEISTER ECKHART, AND *BE!*

The religious scholar Muhyiddin Ibn Arabi, who was born in Andalusia, Spain in 1165 and died in 1240 in Damascus, Syria, underlined the importance of words eight hundred years ago. Muhyiddin Ibn Arabi wrote in his work *Fusus al-Hikam*, one of his masterpieces, that the purpose of each of the prophets and the meaning of their names are mentioned in the Qur'an respectively and are cited in detail. The meaning of the name of the last prophet, Muhammad, according to Arabi is uniqueness. The Prophet Muhammad is a great jihadist who best represents the uniqueness on the path to the singularity in the best possible way. Islam is provision on this path. After the arrival of Islam, other religions didn't disappear. The completion of these religions is within Islam.

There is another Sufi who lived a hundred years after Muhyiddin Ibn al-Arabi. The name of this German religious scholar was Meister Eckhart. Interpreting the Gospel far from traditional dogma, this Sufi still maintained its unquestionable importance today, just like Muhyiddin İbn Arabi. All branches of science have been influenced from the works of these two scholars and it is obvious that they will continue to be influenced. More importantly, with the processing of knowledge at all times and its intensity, knowledge reduces on the timeline and expands on the line of space. How the night and the day flow together are related to this wisdom. Therefore, the idea of Qur'anic mysticism has been in place under the name of quantum physics. From this, it can be concluded that Allah will complete His light.

Religious scholars and books have played an effective role in the birth of physics particularly. Just as how metaphysics influenced the work of Isaac Newton, the founder of physics, quantum physics and the invisible laws of the present have also conquered the hearts of many great physicists from the 20th century until now. This is because, with this theory, the desire to flow towards a singularity has brought itself to the person related to the subject. All the mystery lies in the statement of Shakespeare, *to be or not to be, that is the question.* In other words, Allah's order *Be!* is the occurrence of creativity through impact and the interaction between things and people in the universe. Johann Wolfgang von Goethe, a naturalist from the 18th and 19th centuries who studied German poetry, literary and the theory of color, was given an honorary place in the concepts of creativity and genius. The desire to unravel the secret of the universe was clearly conveyed to the reader in his poems and works.

3.2.11 IS *BE!* POSSIBLE WITHOUT VERSES?

As mentioned in Surah Al-Alaq, the first surah, the core meaning of the Qur'an is to read. This also indicates the purpose of the book, because in the holy book it also means gathering and collecting, and the fact that the verses have been sent in parts and have been completed in different time periods deepens the issue. The word *ayet (verse)* is Arabic. Every letter, word, sentence, surah, or even the *Ayet-i Kerim* as another name for the Qur'an clearly reveals that the pieces or items can be an entirety or a singularity by finding and completing each other.

The semantic meanings of the word verse fall under evidence, open miracles and signs. These verses descending through revelation do not narrow the knowledge in Allah's verses down to the same place. It is necessary to prevent basic meanings from getting confused with the others so as not to narrow down such knowledge. In this way, Allah's words permeate people and things. Hence, the verses descend down on all corners of the city of Medina and Mecca and in every corner of the city. Therefore, the essence of knowledge is located in the bosom of objects and man. Yes, everything is a word because it is only possible to read the universe provided that things are in words. *[And mention] when the angels said, "O Mary, indeed Allah gives you good tidings of a word from Him, whose name will be the Messiah, Jesus, the son of Mary - distinguished in this world and the Hereafter and among those brought near [to Allah]. (Ali Imran 3:45)*

Everything in the universe has a name. When Allah had created the Prophet Adam, He asked him the names of all things in front of the angels. This means that knowledge of objects is inherited by all people. So, why then is there revelation? In a way, revelation is a means of accessing and obtaining knowledge. The revelation of all people is life. It carries every state and all space in the past that was previously treated as space-time and even the absolute hereinafter, and it always speaks of the points of its existence. Moreover, man is a book. Every book that he reads adds different patterns and colors to the very book within him. It makes him mature as an adult and nourishes him. If he were to live in consciousness of Allah, he would paint the universe as a painter with the colors of Allah in his body. *(Say: "Our religion) takes its hue from Allah. And who can give a better hue than Allah. And it is He Whom we worship." (Al-Baqara 2:138)* In his stance is the love of Allah. *And do not walk upon the earth exultantly. Indeed, you will never tear the earth [apart], and you will never reach the mountains in height. (Al-Isra' 17:37)*

Returning back to the topic of books, the names of people and objects are not enough to explain them. However, the more knowledge about a person or an object one learns, the higher the proximity of that person, whose knowledge about these subjects is doubled, increases. With the knowledge of that object or person, he sees more of himself and of them. In fact, it is Allah who is ubiquitous in every place that sees and is seen. On the other hand, the knowledge of anything that sees and is seen takes place in the mind, the heart and the consciousness of the human being. In that, according to the dimensions of the internalized situation, these names transform into sentences and books. With the language of modern science, connotation and its mechanisms are then created.

3.3 THE MIRACLE OF PATIENCE THAT DECELERATES ELECTRONS

3.3.1 TABULA RASA: PAPER, PENS, AND WORDS

For centuries, the term that thinkers focused on was the analysis of life called *tabula rasa*. Although the idea of the tabula rasa extended to the Ancient Greek period, the conceptualization of thought began with John Locke, a 17th century philosopher, who is regarded as the father of liberalism. Although many thinkers did not use the term tabula rasa, they brought different interpretations to the subject.

According to empiricism, the human brain is like an empty slate. And knowledge is acquired through experiences that have been obtained later. According to rationalism, people have some natural knowledge. For example, the Greek philosopher Socrates said, *When the human being is born, all knowledge is already in his head; the goal is to remember it.* For instance, the French philosopher Descartes also agreed with this idea and argued that human thought is innate.

To explain the Qur'an through the perception of quantum mechanics, if everything and everyone is the word, then who or what is the paper and pen? As is known, it is the law of action and reaction in physics that records what is happening in space. This very law becomes both the torch and the fire to the previous question. If the reaction is as much impact as the act of writing, it is the space where this writing act takes place, that is, the paper.

3.3.2 SOLAR SYSTEMS AND REVELATION

TO BECOME FRAGMENTARY QUICKLY

The small letter k tells the small letter y that a new film has arrived at the cinema and speaks about various parts of the film. A war between the words has begun and it will take time for it to end. This is because there are no heroes in the film. The words and their swords are flying through the air. They are too busy getting revenge to smash each other. The disintegrated particles cling to the other letters, which are aligned in order so as not to fall into one, two or three substrates. Some would do so to survive these fragmented forms, some to maintain their mastery and others to admit it. They would even go so far as to show the heartless black parts of the rest of the letters. The winning part was always these letters, and by commercializing the words, moving their places, breaking them up, and using the techniques of cleansing time after time regardless of if necessary, a battle occurred whenever possible.

The small k was affected so much by the film that the particles of the small y suddenly broke apart, and it suddenly turned into the big letter K without waiting for the right moment as it was talking about the film.

In the case of *Schrödinger's cat*, the assumptions such as whether the cat is dead or even aware of its own presence will only develop according to its follow up, i.e., by the choice or decision of the person watching. Therefore, the result will be reached depending on the vocabulary of the person watching. So words also have a demand. They want awareness. This should be such awareness that the conception of the cat regarding the fact that it has been locked into the box and abandoned to danger cannot be dominated by the authorities who have information on the issue. This situation is only solved in the human mind as a short-term assumption.

The most work here falls to the people who work with words one by one. The infinite particles of responsibility are given to all people in the universe or to the carriers of all concrete and abstract concepts. These are the hereafter, the feast, the solar system, in short, Allah's creations, and all of them need revelation so that these perfect operations and processes can occur. On this basis, revelation may come from Allah or the angels who are thus appointed. In fact, a revelation exists within every word and all knowledge (good and evil). The important thing is to see how much faith this knowledge contains. In this way, the book that announces and informs human beings and objects in the universe also delivers the uniqueness of integrity to the receiver of the revelation. We are in need of Allah and revelation. Otherwise, we wouldn't have been created with a nafs. Thanks to our nafs, we are surrounded by two values and the sun of the truth rises here. Just as we need it, solar systems also need revelation. The music of the universe is only thus transformed into the Yellow Horn in our ears. Just as life has a duty, so do those who live in it. The basic task is the same for all creatures: Obedience to Allah and faith performed consciously or unconsciously (even if one does not believe in Allah, His breath does). According to this infinite form of obedience and faith, each object or man has a mission.

Every object or man principally has a group of primary duties. For example, the duty of man is to be a man, the duty of an angel is to be an angel, and the duty of the sun is to be the sun. In other words, whatever the object represents, it will perform its own existence. According to the correlation principle everyone and everything serves as helpers of other objects through visible and invisible duties in relation to their own mass, energy and mission. This occurs only when the substance or the body is transformed into knowledge in proportion to the existing but variable potential of the object or the man.

According to the terminology of time, everything is made up of moments. But awareness of moments is necessary, and the extent to which it is visible corresponds to the dimension of knowledge with regards to the object. According to the string theory, awareness is always permanent because very low level of awareness sufficiently contributes to human life. At the same time, it regulates the systematic dimensions of the inter-universal relativity theory. For example, it is a system that shows how a mathematic principle works between our seventh universe, which equals a thousand years, and the time period that corresponds to a day. Time being far or near is different based on the interaction of knowledge between the object and the human, i.e., it is relative.

3.3.3 GABRIEL, THE ANGEL OF REVELATION, AND THE STAR SIRIUS

Allah taught the Prophet Adam all the names of the objects. He provided blessings for man both on earth, in the heavens and in his soul. Thus, it is imperative to understand the dimensions, words and concepts in objects so that they can understand and fulfill their humanly responsibilities. Allah gave Gabriel (PBUH) many duties. They are his secondary duties, and by the order of Allah and according to the burden of knowledge that he could carry, he is equipped with the qualifications of coordinating some of those universes with the angels and, when necessary, becoming water for the sun. In short, if needs, He assists and accompanies the angel of nature, Michael (PBUH), or vice versa. The fundamental importance of Gabriel (PBUH) is that Allah glorifies him as an angel of revelation. Because Gabriel (PBUH) brings this revelation, the knowledge of this revelation is very large and deep.

The meaning of revelation here refers to the hope that it brings, and this hope tells everything regarding it, namely the apocalypse, the hereafter and the heavens. Each word corresponds to a star, a group of stars, etc. in the solar system. This hope also has its own star. It is the brightest star called Sirius. When Gabriel (PBUH) appeared to the prophets, he came in human form and this star came down to earth. Could it be that the star Sirius is the intersection point of the revelations that spread throughout the universe? This is because Gabriel (PBUH) passes through Sirius stars, defined as seven ABCs through the path of Fatiha, while descending to the first layer of the universe from the Sidrat al-Muntaha located on the seventh floor opposite from us. Just like the years of the universes that are located seven floors down decrease, the dimension of Gabriel (PBUH) changes shape and shrinks equally, yet his shadow stays in the universes and maintains its size. In a similar vein, the servant also needs to pass to the Sidrat al-Muntaha from the earth through the Fatiha. The servant, who is aware of himself, knows which star belongs to him.

as the example of an innocent child

sitting
hoping

the sofa, door, lamp,
and arriving guests
are not be able to see it

mediocrity
has made my heart
forget its presence

it's left
with a festival
when the day turned into night

it didn't even leave me
its shadow
in the rooms of the house

it is the

star of Sirius
the owner of vows

3.3.4 IS THE DARKNESS MATHEMATICAL?

We have made the Night and the Day as two (of Our) Signs: the Sign of the Night have We made dark, while the Sign of the Day We have made bright; that ye may seek bounty from your Lord, and that ye may know the number and count of the years: all things have We explained in detail. (Al-Isra' 17:12)

Is the darkness mathematical? Toanswer this question, it is necessary to understand both darkness and mathematics. As in everything else, the darkness and mathematics are words. However, just like light, darkness has also many faces. How is it that all of these faces are calculated as something invisible?

3.3.5 THE SEVEN SKIES

Did I dream I was a butterfly,

Or was it a butterfly which dreamt it was me?

Chuang-Tzu

Today's cosmonauts are not only interested in the structure and history of our universe, they pay the same attention to other universes too. Although it is difficult to comprehend, these universes do not necessarily have to be real, it will be enough if they can be explained by physical laws. In particular, the ideas that regulate this *real* idea need to be examined. Perhaps, the main theme word would be united under a universal structure through the expressions of *reality*, *accuracy* and *truth*. So much so that, in the dualism in which man exists and therefore can live, there are worlds and their absolute truths.

These truths differ from person to person. After all, isn't the individual different? The difference is proportional to the awareness of one's truths. These awareness levels also have atomic steps that can be reached in terms of knowledge. These steps, which differ from the structuring of dualism, have two conditions instead of accepting them as they are and imagining their future with errors. These steps are to learn and the act of asking to learn.

Of course, we are equipped with the potential to inspect the solar system. But other than this knowledge, if vital features gradually decrease, then there will also be deviations in knowledge. This can be called information pollution. This kind of knowledge unfortunately stabilizes the balanced functioning of the universe in a negative way. Nevertheless, the universe also has its duties except for human. Each universe has a distinct characteristic as a form of self-expression because any feature is a responsibility for and therefore a duty to a mission.

All these universes are in harmony with the laws of nature despite their differences in the integrative systemization. This is because the nature does not stand back from being available to people through Allah's commands. To assume this as a coincidence contradicts the duality of gravity and the principle of relativity. And here, it will not be possible for both physics concepts to create their own language of truth.

3.3.6 OBJECTS AND THEIR SHADOWS

All that Allah has created will prostrate themselves before Him. This includes the shadows of objects too. These shadows diminish and turn right and left, then, prostrate themselves before Allah. Those who are in the heavens such as the shadows, the living things on earth, and all the angels prostrate themselves before Allah without patronizing.

3.3.7 THE PROSTRATION OF OBJECTS ACCORDING TO THEIR CREATION

Allah's command has come. Allah says, *Do not hurry to ask for order, even though it is there.* Instead, through His command, Allah sends angels to announce Him through His revelation. All the authority of *Be!* has a purpose that leads to an end, and the end is resurrection. In other words, the true promise Allah has made to Himself is that He shall raise people up again, and there will be an absolute result. Allah's command for them to see the truth is *Be!*

Allah punishes evil. His punishment is good because it includes His affection and mercy. Allah punishes the evil so that humans take their shadows as examples. The shadows shrink and prostrate themselves before Allah. The way of prostration varies from one object to another. In other words, according to its shape, the object prostrates itself before the Creator.

And if all the trees on earth were pens and the ocean (were ink), with seven oceans behind it to add to its (supply), yet would not the Words of Allah be exhausted (in the writing): for Allah is Exalted in Power, Full of Wisdom. (Luqman 31:27) It is obvious that there are people who prostrate themselves according to their personality, knowledge, in short, their characteristics. Moreover, the same person can never make the same prayers. All creations in heaven and on earth prostrate before Allah *without patronizing.* They adhere to the commandment, and Allah's order is amalgamation. To live and to die, everything is full of sustenance. And Allah is the creator of all sustenance. But those who are stingy deny that Allah gives sustenance. Stinginess refers to bad attitudes and behaviors. Those who do not believe in the hereafter have a bad name. Allah owns the mightiest of names.

The Qur'an is an explanation, guidance, mercy and good news for everything. But most people don't know this. They don't even make any effort to know it. However, knowledge is very important. No verse sent by Allah disappears. It only changes according to the experience of the person and his life, in short, his knowledge. The truth of the verses gains degrees in terms of color and size according to his comprehension of faith in Allah as well as His names and titles. The extension of these degrees even goes to heaven. This leads to reversible degrees; thus, the formation of different combinations of objects becomes reinforced. Revealing the truth, or more precisely, the positive atomic flow and rotations, which are derived from the way of human life, purify human beings from their sins. Should Allah has granted him *perfect* repentance, he is then given the ability to turn sinful wine into a rose, as is known in the case of Rumi.

Of course, the infidels were also given the perfect creation and its prosperity to reveal this knowledge. In consequence, Allah has blessed all people with his soul. A person who rejects faith in Allah mentions Him in every breath he takes. This faith coincides with what exists in man in spite of all evil. But the important thing is to make faith visible. Such human beings have the hope of rising to the heavens. This carries them to the only knowledge, the knowledge of Allah.

3.3.8 THE LANGUAGE OF HEAVEN

There is a reason for the creation of the day and night just like everything else. The main theme of creation is the determination of faith and good behaviors. A particularly daunting task is given to the night and the day in this regard. For Allah's sake, it is important how many good manners one displays, how one does jihad, how one is grateful, and what one does for mankind. In short, it is important to know how much the servant is aware of the moment and how much he is grateful. The most distinctive characteristics of people who realize the moment are modesty and tranquility. A man of such personality does not chase after evil, and Allah turns his evil into good. He is calm and kind. He is a generous person who loves to share what he has.

In fact, there is a *greeting* in every beautiful work that a gentle person does. *Salaam* is the language of heaven. The sole thing the souls of heaven speak of is *salaam*. This is because these people spend the night prostrating themselves. They leave their values, repentance and grace to the night. They say the word *salaam* means that they have reached the knowledge of patience. People who reach the knowledge of patience and turn evil into good will reach the highest level of heaven because the homeland of one who sustains eternity, and lives for it, is also eternity. It is the house of the servant that he built in pure love and truth.

What is the primary element of living out pure love and the truth? The Surah Al-Furqan (25:70-77) gives a detailed and complete answer to this. The primary element of truth is repentance and faith. If the servants repent to Allah, they will escape from evil. They shall be protected from it, and in return for their jihad, Allah shall place faith in their hearts in small bits. Thus, they won't testify to a lie. They shall always be obedient to the verses of their Allah. They shall plead to Him for a blessed spouse and their lineage. Their attitudes and behaviors shall be the work of patience, and Allah declares that He will bestow the highest authority of paradise upon them for their patience, and that a ceremony waits to welcome them with respect and greetings.

Briefly, if there were no penitence, that is *Say, "What would my Lord care for you if not for your supplication?" For you [disbelievers] have denied, so your denial is going to be adherent. (Al-Furqan 25:77)*. Therefore, the last verse of the Surah Al-

Furqan means that just like the place of the gentle people (Illiyin) and the place of those who desire that evil is apparent, the homeland of those who do not realize the place of Allah (Hutamah) is also apparent.

3.3.9 PRAYERS AND BALANCE

To believe in Allah leads one to remember Him. One remembers the promise he has made to Allah (the world of spirits). This helps people get out of the darkness. *Do not treat Allah's Signs as a jest, but solemnly rehearse Allah's favors on you, and the fact that He sent down to you the Book and Wisdom, for your instruction... (Al-Baqara 2:231)* Reading the Qur'an is to remember what happened in the spiritual world. Allah's response to prayers is already confirmed as much as the servants remember those in the spiritual world.

There are three types of accepted prayers. The first is based on the name of the Most Gracious.

The fulfillment of Allah's commandments is a response to peoples' demands. The second is the transition from the name of the Most Gracious to the Most Merciful. Here, as the servant fulfills Allah's commandments, Allah accepts the servant's prayers. The last one is that man must fulfill the commands of Allah so that He shall reply to His servant with the names of the Most Merciful, the Truth and the Guide...

3.3.10 RESPONDING TO PRAYER

Only Allah accepts prayer. He is the only one who ensures balance in the universe. And it is only He who takes us from the darkness into the light. There is a creature in the darkness that adversely affects the light. He is the Beast of the Earth as mentioned in the Holy Qur'an. He lives underground. He always wants to undermine human beings by mixing in with the force of gravity where he is nested and influencing their balance on earth through the form of the geometric masses. This creature, which draws man to evil with his power, also informs Allah of those who do not have faith in himself. Allah has also made the evil a witness like the beauty of faith. As those who serve evil build their own path to the place of judgment, the road approaches them, becomes clearer and Allah's warnings about worshiping evil become inevitable. However, salvation and the good news come from Allah only and denial is the easiest way that people can follow without comprehending the owner of the universe, Allah, and His verses. However, if they would think for just a moment, they would see that *mountains move like clouds.*

The situation mentioned here is the clear and implicit mobility (according to Allah's names and titles, the Manifest and the Hidden, or in scientific terms, physics and quantum physics). Therefore, the aims of these two objects are the same but their tasks that lead to the goal are different.

There is parallel visible activity, but if it were to be explored, there would be visible mobility **in the future.** (bunu ekledim daha açıklayıcı olduğundan)Because their velocity (velocity of light) corresponds to the same ratio with respect to their mass, their equality with their energy remains constant, and vice versa (Einstein's theory of relativity: $E = mc$; $E = mc2$). In fact, they do not ensure stability by their own speeds, masses and energies only. They benefit from multi-purpose asymmetric structure arising from action and reaction states formed by other visible and invisible objects. This is because one of the functional tasks of the asymmetric structure is to be active in the establishment of the balance by constantly renewing itself with the *exchange system* by including heterogeneous objects for being necessary in another homogeneous or even equilibrium method. This system must be fair. Otherwise, the structure of the universe goes wrong. The sky and the earth nest together. The

Almighty Allah owns all the dimensions of the world. With His commandment, natural disasters and deaths occur for the order of justice.

Another aim of the asymmetric structure of the object is to protect the particular balances of each object characterized by the same or similar tasks within themselves as much as they can and to assign their weight, i.e., excessive energy, to a certain target collectively. Since the target does not respond to such electrically charged density, it also brings the bottom to the top so that it can regain its balance. For example, there may be earthquakes in a certain location. With a very short physics definition, if the number of protons in the atom is equal to the number of electrons, the atom is unloaded in terms of electricity. But if these numbers are not equal, this particle is an ion. The structures of these ions are highly unstable and interact with other ions and atoms in the environment to avoid their high energy.

3.3.11 PATIENCE AND TIME

Allah informed the last prophet, Muhammad, about the twenty-five prophets and their verses as well as the wise men and believers by repetitions in various ways. *And you, [O Muhammad], were not on the western side [of the mount] when We revealed to Moses the command, and you were not among the witnesses [to that]. But We produced [many] generations [after Moses], and prolonged was their duration. And you were not a resident among the people of Madyan, reciting to them Our verses, but We were senders [of this message]. And you were not at the side of the mount when We called [Moses] but [were sent] as a mercy from your Lord to warn a people to whom no warner had come before you that they might be reminded. (Al-Qasas 28:44-46)*

The Prophet Muhammad had learned about the prophets, their lives, their suffering, their struggles on the path of Allah, and their different qualities through revelation. With the verses sent down by Allah, the divine light of faith in the heart of the Prophet reached Allah. *So did (Allah) convey the inspiration to His Servant- (conveyed) what He (meant) to convey. The (Prophet's) (mind and) heart in no way falsified that which he saw. (An-Najm 53:10, 11)*

Every prophet is a servant, while some are thought to be both a prophet and a messenger. Allah asked some of them to fulfill their mission according to their particular characteristics. With the touch of that single direction to the endless different dimensions, Allah will have completed His light. The same is valid for an ordinary servant. Living in this life with the knowledge of the creator is hegira. Withstanding all illnesses and sicknesses during the hegira is only possible through patience. It takes people to the truth. Patience is within the time and the simplest color of time. It is yellow, and patience is the crown and glory of beauty to give direction and strengthen all emotions. The servant tells the names of Allah in the most beautiful way through patience and this narration is the voice of water.

3.3.12 THE MIRACLE OF PATIENCE WHICH DECELERATES THE SPEED OF ELECTRONS

It is patience that allows electrons to stay in orbit and reduce their speed or that stabilizes them. With this feature, patience is a vital factor in the formation of infinite shapes by forming the edges of geometric shapes. Every man must be patient because patience has an infinite form and is therefore the main factor in shaping all other features. But when one begins to shape patience with faith, he introduces love to any geometric shapes known to them on their own edge lines. These geometric dimensions change, and are transformed and shapes become pliable through the awareness of truth. The most common consequence of the pliability of all components is the reduction of becoming asymmetric or the asymmetrical behavior. As a result of these events, the truth comes into existence or occurs in place of where dualism is dense.

4 THE PARADIGMS OF QUANTUM MECHANICS

4.1 M-THEORY AND FATE

4.1.1 SCHRÖDINGER'S CAT

like Schrödinger's cat,
my dreams are imprisoned in a box.
maybe not even a box.
if there were no box, how could I exist then?
perhaps,
there is no me, no cat and no box!
but what about dreams?
I hope they exist.
sitting here,
I have been telling myself about me;

*the me who is deprived of the cold and the hot.
in fact, neither am I myself
nor is someone else me.
neither does it want to seek so as to find
nor does it want to seek as to lose again
now it is too late
even though there is something to do
do not do anything
if there is anything to do
do not even imagine the cat
or else what you will do
will one day take place*

In 1935, the Austrian physicist Erwin Schrödinger relayed the idea of quantum physics to the microcosm to make it visible from the macro cosmos universe. At the center of this thought experiment there was a cat, a box and a particle that was likely to deteriorate. The living cat was in the box and so was the particle. The box was closed and the condition was to keep it unknown what would happen inside it. Only estimations could be made. If the cat were to break the particle, the gas would spread throughout the box and the cat would die, or if it were not to break the particle, it would remain alive. This addresses the idea of dualism. However, if the box were to have been open, the condition of the cat would have been revealed. However, as long as the box is kept closed, there would be no knowledge about the cat. The most important feature in this experiment is the lack of the phases of time that would not be observed in the light of time.

For example, let's say that the cat is dead... So, knowledge of how it has died is very deficient. Did the cat reach its death from the right, middle, or left of the box? The beginning and the end are the only things able to be known. Knowing the end depends on the decision to open or keep the box closed. This is called the super position principle. The most impressive feature of this principle is that there is only one truth. Although nothing seems to have happened, the case is the opposite. Another truth can be derived from here that is the egalitarian principle, which goes towards the whole or arises from it.

However unusual it sounds, if we do not include the concrete reality of Schrödinger's equation, the universe we live in cannot be depicted. It should be noted that there are limitless human forms and lives that experience every different dimension, transforming possibilities into experiences. Many books could be mentioned here from an artistic perspective. However, the protagonist in Hermann Hesse's *Steppenwolf*, published in 1927, could be shown as an appropriate example in the sense that as a human-animal, referring to the fluctuations he experiences due to his wolf-like characteristics, he displays distant attitudes from what a homogenous individual would present. Hence, probabilities are instructive elements that prevent people from falling into gaps and it is obvious that they address each person differently.

However, there is a very important and quite simple matter to consider. In Saint Exupery's work Little Prince, the pilot draws a picture of a few lambs for the Little Prince. Yet none of these pictures appeals to the prince. At last, he draws a box with holes in it stating that the lamb is in the box. The Little Prince becomes very happy. There is a little but significant difference between the Little Prince's and Schrödinger's box. Just as the need for possibilities is eliminated by putting holes in the box, other inevitable probabilities are also prevented. Thus, whether there is a cat or a lamb in the box, our will, which resembles the box, learns how to breathe properly. It conceives the meaning of the truth and most importantly, in this case, the probability of killing, which occurs through the insolubility of the possibilities, is eliminated. Living a life as clear as running water...

4.1.2 AN INTERPRETATION OF EWG/MANY-WORLDS

Schrödinger's cat has become the starting point for the Interpretation of Many-Worlds (1957; Everett/Wheeler/Graham), one of the interpretations of quantum mechanics. According to this theory, as we are only in a certain position, we perceive only that state. Therefore, the super position is correct. However, it describes the realization of the results considered as probability. For this, there must be countless parallel worlds. The case with the example of the concrete Schrödinger is that when the box is opened, we will know whether we live in a parallel universe in which the cat is alive or dead. In this way, all binary values get actualized simultaneously alive or dead.

The example of the television will make this clearer. Let's say you're sitting in the living room and watching a particular TV channel. At the same time, something is taking place on other channels. The multi-world interpretation deals with the attainment of the atomic density of infinite emotion and the activation of the infinite variety of gravitational force. This variety attracts the necessary parts for other formations. The universe is directed according to the unity principle and the balance of the objects. This is made possible by the fall of gravity, which is the closest to emotions, and the proportional magnetic weights of such, as well as their recognition features of that emotion, on all the visible and (yet) invisible space time continuum, by means of emitting and absorbing that emotion. Thus, through the integration of motion and mobility, balance is established. The concepts of motion and mobility are related to humanity and truth. The former is stable, i.e., less mobile, while the latter is much more animate. However, a balance between both is necessary so that a universal unity can be attained.

When a person falls into the same space-time continuum with small-scale differences, which are similar to the ones that he repeats most, or asymmetrical, a path occurs from other paths that shape the life of the person. By transforming these paths into a constant state, the person determines his own destiny and thus, that of the universe. The likelihood of repetition of the numerical occurrences of fate is proportional to the knowledge of dualism in the existing, apparent particles of expression. This knowledge corresponds to good and evil just like the two sides of a coin. According to the theory of relativity, such dualistic expressions respond in an encrypted way in parallel with the two-valued data on the regression line according to the constant and invariance principle, in spite of the volume differences in the cosmos. Since the truth is presented to the universe through electrons, both the components of truth and its illusion mix with the effect that it creates and the reaction from it. As Allah says, *They are the ones who have lost their own souls: and the (fancies) they forged have left them in the lurch! (Hud 11:21)*

4.1.3 INTERNAL-EXTERNAL MULTI-WORLD THEORY

Let's say a person is ill, feeling uneasiness in his stomach. Chronic gastric discomfort results in the unexpected and simultaneous pain of the organs, perhaps even damaging them for a short period of time due to the nervous system. In other words, we are in a world in which the most painful diseases exist, and the pain takes shape according to our worlds. This also affects the external world of the sick person, which results in not doing duties properly if we go to work, or never being able to go to work. In fact, if we decide which of them will be our world with the choices increasing in this direction, it takes the real direction and shape of our lives. The patient's choice is of course limited due to his illness. This choice also depends on his resistance and age.

4.1.4 M-THEORY AND FATE

you've badgered fate
your hands are burning
but still you cannot take them away
because
you don't understand anything
because
you are too busy
trying to explain

fate
where are you going
why are you picking on me
stop and look, I am always falling
these roads are so unfamiliar to me
let go of my arms, I shall go back
or at the end of these roads
death may
be waiting for me

Quantum mechanics is actually the soul. Therefore, the unification of information is only possible with the spirit and/or fulfillment of the necessity of expressing the objects according to their determination in the very moment of determination. In these worlds, everyone takes the lead according to their preferences. Everyone has separate parallel worlds. Although these are different to everyone and they have different effects on every person, the worlds they decide on together and jointly are distributed in parallel worlds as waves and particles. Each wave has a field of view that results from the chemical components of each particle. This field of view is the reality that leads to the truth, or it is the measurements that Allah has determined. *The sight [of the Prophet] did not swerve, nor did it transgress [its limit]. (An-Najm 53:17)*

4.1.5 M-THEORY AND (SUPER) STRING

The dream of theoretical physicists in the 20th century is to be able to explain both classical physics and quantum physics under the same roof. In the 21st century, this secret system was opened up with the so-called string theory. String theory is a modern branch of physics that has succeeded in combining the microcosm of the macro cosmos in Niels Bohr's quantum mechanics experiments and Einstein's theory of relativity. Allah says in the exalted book that the mountains move like clouds, but man supposes that they stand still. One of the names of Allah is the Hidden (invisible) and the other is the Manifest (visible). The important thing here is what the word Hidden means or what it doesn't mean.

Hidden, in a way, is the yet unknown. It is the dark one. As in all the names and titles of Allah, the Hidden bears knowledge that is not available to man and which man cannot reach. Allah forbids them to chase after this knowledge. *It is*

He who has sent down to you, [O Muhammad], the Book; in it are verses [that are] precise - they are the foundation of the Book - and others unspecific. As for those in whose hearts is deviation [from truth], they will follow that of it which is unspecific, seeking discord and seeking an interpretation [suitable to them]. And no one knows its [true] interpretation except Allah. But those firm in knowledge say, "We believe in it. All [of it] is from our Lord." And no one will be reminded except those of understanding. (Ali Imran 3:7) There are also those who do not appear in this context as well as visible things. It is not surprising that knowledge about dark matter is limited.

4.1.6 MALDACENA AND ADS/CFT APPLICATION

The most important contribution of South American Juan Maldacena to the world of science was the ADS/CFT application. Maldacena successfully explained the functional association between this holographic principle and Einstein's *Theory of General Relativity* and quantum theory. Maldacena, in the Anti-de-Sitter (AdS) space, aimed to reach string theory and the equivalent of the conformal field theory. In plain language, it meant to try to express the material-spiritual dimension and harmony between good, evil and the truth. In other words, in the field of theoretical physics, the geometric shape does not undergo a change in terms of angle when the particles have high degrees of symmetry during interaction.

4.1.7 THE BREATH OF ALLAH-1 MATRIX

when eyes talk
about what has been told
who listens to them
except the ears

what is known is right here
but what else can reach
the skies but the heart.
other than the heart what can reach it

the unbelievable
is in the yellow of the light
a ton of love
flows from them into the senses

isn't it our creation
the seven verses for the Fatiha
from the seven layers of the universe
and the mother's womb

while our skies are the same
how could they
not care about
the thing that makes us who we are

Allah's breath is but one. The cosmos and humans are fed by this breath. And one breath takes two values as the exhaling and inhaling of the universe. In other words, the word that comes from Allah, wants to contemplate and surrender spiritually, maintains the functioning of the universe systematically, and is the Fatiha of Allah's names and titles, makes objects submit to not a necessary shape but endless change that is visible or invisible in every breath. But the breath of the universe solely responds to the truth, which it takes from the breath. This is because the path from our solar system to the seventh universe is less asymmetrical. This means that human beings in the lowest layer of the universe experience the truth; however, they are led to a distance seven times further, in other words, to the extent of the comprehension of the seven verses of the Fatiha. The asymmetrical structure is replaced by symmetrical structure.

That explains how the Prophet Muhammad could see Allah when he ascended to the miraj. Vice versa, the places of a man, who does not execute his Fatiha, increase, giving rise to Taghut. This person deprives himself of the knowledge of Allah. Here is a generalization: However, you see the world is also how the world sees you. This is because you have chosen the very feeling you are feeling at the moment. And your eyes are showing you exactly what you have chosen. *His command is only when He intends a thing that He says to it, "Be," and it is. (Ya Seen 36:82)*

4.1.8 ALLAH'S BREATH-2 (SUPER) STRING

And hold firmly to the rope of Allah all together and do not become divided. And remember the favor of Allah upon you - when you were enemies and He brought your hearts together and you became, by His favor, brothers. And you were on the edge of a pit of the Fire, and He saved you from it. Thus does Allah make clear to you His verses that you may be guided. (Ali Imran 3:103)

Allah is in the seven level universe. He is aware of us in every step we take as a whole. This principle of wholeness takes both from here and from what we leave to the universe that we have within us, and it takes even more knowledge from us all over the universe, as well as from us that flows from the universe.

[And Luqman said], "O my son, indeed if wrong should be the weight of a mustard seed and should be within a rock or [anywhere] in the heavens or in the earth, Allah will bring it forth. Indeed, Allah is Subtle and Acquainted. (Luqman 31:16)

Allah is aware of what is happening in the soul, heart and mind of our bodies in every step we take, as well as the role of the soul and Satan that affect them. This is the fate of their quantum. Together they write the fate of the person by combining the particles of past and future destiny with the nature of the power of gravity. The fate of the person is the hereafter. Allah will forgive us for the flaws of our human attitudes and behaviors that we have taken to construct our own hereafter. The beginning of this is the fact that human beings are aware of what they lack of, knowing that they can improve through repentance to Allah. Adherence to Allah's rope is necessary in every sense. Otherwise, man abandons all the Fatihas given to him innately to a black hole. In other words, acts of worship from the cause relations of material domination become indispensable to man. Thus, even though numerically equivalent, seven layers of electrons cannot be controlled by protons during mobility, and symmetry and asymmetry can be controlled by entropy.

As in all entropies, the shades that emerge in these colors act as a signal and activate the surrounding magnetism according to the mass density numerical data carried by the colors. Here, Allah bestows His servant who absolutely needs Him to save himself from artificial worlds, including those that have established their systems or haven't established any (hallucinations in a sense):

And say, "My Lord, cause me to enter a sound entrance and to exit a sound exit and grant me from Yourself a supporting authority." (Al-Isra' 17:80)

Otherwise, people worship in their own way (idols) in multiple worlds. They deify them and thus deviate from the goal of worship. However, Allah created man and all of his activities. There are no intercessors of worshipers. Instead, those who utter the aforementioned prayer touch as much truth as possible from multiple worlds. As they do so, they enable the holy light to spread from the particles of truth into the universe and from here to the seven levels where they rise and fall. Those who persist and those who believe will be appointed by Allah to guide their societies. Allah gives the right of being intercessors to His servants who have the pure features of the

Prophetess Mary. This will be in the presence of the Most Gracious. Allah provides sustenance to all human beings, even those who do not believe, through the name of the Most Gracious. Therefore, being an intercessor in the eyes of the Most Gracious means that Allah loves this servant.

4.1.9 THE GOOD AND THE BAD

Beautified for people is the love of that which they desire - of women and sons, heaped-up sums of gold and silver, fine branded horses, and cattle and tilled land. That is the enjoyment of worldly life, but Allah has with Him the best return. (Ali Imran 3:14) It is clear in the above verse that the nafs has electromagnetic properties due to its atomic characteristics. Thanks to the functional properties of this magnetic feature, dualism comes to life in the nafs. But magnetism has its own field of vision also, which still forms its own.

It is the right or the rights of truth. *In it are clear signs [such as] the standing place of Abraham. And whoever enters it shall be safe. And [due] to Allah from the people is a pilgrimage to the House - for whoever is able to find thereto a way. But whoever disbelieves - then indeed, Allah is free from need of the worlds. (Ali Imran 3:97)*

There's a truth. The truth is the verses of Allah. Thus, Allah does not act unjustly. Allah creates man in the space-time continuum in accordance with a good-evil dualistic character. He perceives it in this prudence, and from here, it becomes his truths and sins. Allah does so to determine faith and purify sins caused by the duality of His servant. The invisible rope of Allah has already tied up the universe, and whenever the faith of the servant increases, he firmly attaches to the rope of Allah.

Those who hang on the rope of Allah will fear Him. They know that the good and the bad come from Allah. Being thankful becomes the oxygen of the heart. Avoiding what Allah dislikes and rebellious attitudes and behaviors means to be thankful to Allah and to express gratitude in contemplation. Again, fearing Allah is only possible through patience; and patience is stable but not conducive to deviations. Stability remains in essence. In the world of essence, the servant sees the verses in the universe. He sees the truth only in the verses most intensely. From there, the servant approaches to Allah as he sees the angels, the prophet, the books of Allah, and goodness and evil. Therefore, despair and sadness cause perseverance and faith to weaken. Allah has not created the causes in vain. Now that human beings live in dualism and their nafs is created in harmony with both the revelation of truth from the universe and the permeation of the superstition to the universe, the best example of who used the reasons in dualism as a way of logic is the Prophet Dhul-Qarnayn. The starting point of these reasons is of course good and evil and solving logically this dualism, that is, such values as win-lose and rise-fall far from value judgments leads human beings to the place they need to reach. The purpose of Allah in creating good and evil was to bring the above-mentioned essence into the light and out of the darkness. Light is faith. Allah puts the good, evil and the varying faces into the day so that faith may be determined and that He may testify among people. In addition to this, the object responds to what it does and what people do in their unique way.

The sins of believers are only purified through their faith. Those who do not believe deserve to be destroyed. No one can get to heaven without revealing those who are on jihad. And when a person makes a mistake, he doesn't die without paying the price for his mistake. The important thing is that one wants forgiveness from Allah and hopes that Allah will forgive him.

Those who fought in the Battle of Uhud together with the Prophet Muhammad made mistakes and left their places instead of staying in the hills where they were appointed. Allah gave them grief upon grief. In such situation, the servant who cannot even feel the pain within such grief struggles in confusion; in a sense, he experiences his own doomsday. Then, Allah forgives him and puts trust in His servant. While the Prophet Jonah was charged with prophesying to more than a hundred thousand people, he abandoned his place due to people's disrespect for His word. However, success or victory over the enemy as well as patience and perseverance is possible. The person who strengthens these factors will be prepared and alert to any situation he may encounter. He fears Allah. In the Qur'an, the oppression of the prophets and the cruelty suffered by the enemy during the communion of Allah is frequently mentioned. The wrongdoers are the rulers of these times and the people who had a voice.

4.1.10 THOSE WHO SOW DISCORD

And they followed [instead] what the devils had recited during the reign of Solomon. It was not Solomon who disbelieved, but the devils disbelieved, teaching people magic and that which was revealed to the two angels at Babylon, Harut and Marut. But the two angels do not teach anyone unless they say, "We are a trial, so do not disbelieve [by practicing magic]." And [yet] they learn from them that by which they cause separation between a man and his wife. But they do not harm anyone through it except by permission of Allah. And the people learn what harms them and does not benefit them. But the Children of Israel certainly knew that whoever purchased the magic would not have in the Hereafter any share. And wretched is that for which they sold themselves, if they only knew. (Al-Baqara 2:102)

There has always been someone in opposition to those who sow discord among people. The Prophet David killed Goliath. The Prophet Moses fought against Pharaoh. The Prophet Muhammad (PBUH) fought against Abu Jahl, that is to say, against all ignorance. It seems that the good and the right have always defeated the evil and this will continue to be so.

On the other hand, the evil of some people is being wiped out by others' evil. Otherwise, the earth would be turned upside down. This explains the principles and systematic operation of the divine order. In other words, one aspect of ensuring the balance of good-evil is the destruction of evil. The rebellious people who insist on evil push themselves into nothingness by disregarding Allah's grace. What awaits these weak people, who are quickly caught up in enthusiasm, is expressed by the last pure and specific words. Being a powerless existence proportionate to their age and structure, based on their age and structure, it is always the misfortune of people who have embraced evil because they cannot stand behind something nor can they desire such a thing.

On the other hand, instead of sowing discord among people, they explain the meaning of life with the faith and patience because they make sense of Allah's names and titles. For this reason, it is necessary to understand and conceptualize time to respond to the tendency of all matter and waves that constitute the structure of humans and the universe. This means that the person can affix and direct his own attitudes and behaviors to the face of internal and external factors. For this, one must explore the realms of knowledge and concepts, be present there, and from there control his actions as much as possible. To do such a thing, it is necessary to know their existence, for example, how a substance emits an energy or radioactive dissolution into the atmosphere during its transformation to another state.

According to scientific data, in particular the yellow color and its contributions should be examined under a different lens. Is the yellow color the particles of Raphael's horn spreading into the universe? Raphael waits for the command of Allah to blow into it. He patiently waits for the end of the horn sound for the universe to be completed. Is the sound of the last horn the call to reveal the names and titles of Allah?

4.1.11 GOODNESS-EVIL/BIVALENCY

But if the Truth had followed their inclinations, the heavens and the earth and whoever is in them would have been ruined. Rather, We have brought them their message, but they, from their message, are turning away. (Al-Mu'minoon 23:71)

Whatever Allah has bestowed upon man and whatever is manifested in man is his welfare.

Allah's favor towards him is grace. This favor bestowed upon man is never to be imprisoned in him. It has been bestowed to be given to others. Otherwise, people would be miserable. Therefore, if a person has knowledge or skills on a subject, he should open it up to the public. This is his knowledge. In the same way, the rich will give their alms because a person can reach the hereafter through sharing. Truth arises from the light of goodness. Its light signal comes together with his somewhere and they emit light waves from the place where they gather. If the given blessings are not transformed into goodness, there will be no good in those blessings. They become worshipped. Allah does not oppress those who do so; rather, they bring upon themselves their own destruction.

Some people worship idols, and this prevents them from finding the right way. The right way is to serve Allah. Pharaoh's servant Qarun can be shown here as an example. Allah gave him knowledge. With this knowledge, he became rich. He considered himself to be the owner of properties and the right owner, and by having such an idea, he ultimately denied the hereafter. However, that knowledge was not given to him to combine it with self-seeking artificial manners. Nor was it given to spread discrimination and evil. His duty was to serve humanity through this knowledge. No matter what, it was his test to decide whether or not to share it. This is because one dimension of jihad is material and spiritual alms.

The names of the principle of forgiving evil are also faith and good works. People who have adopted this lifestyle are candidates to take part in the good community. Moreover, good persons are not alone; Allah is always with them. This is because where there is beauty, there is definitely Allah who holds all truth. Evil is also the deception of Satan who invites us to do evil. Satan doesn't leave the man's side until he does evil. When he succeeds, he lets the person down by *saying I fear Allah.* And as the things they worship shall escape from the person in the hereafter, so will Satan.

Returning to the subject of sustenance, the conclusion of provision is that the eternal provision, which Allah offers to His servants, increases the vulgarity of those who oppose His commands. And those who obey his commands increase their faith. The sustenance is forked into two directions. One side is the world and the other is the hereafter. Every man builds his own hereafter and paradise within his worldly life. As the number of the idols human beings worship increase, worshiping money and status also becomes prevalent. In that, one makes friends with all but Allah. Allah thinks this situation resembles a spider's web. The web of the spider is feeble, unstable, and disappears in a blink of the eye. Another issue here is that the spider uses its web as a trap; it always seeks prey and catches its prey with that web. The spider gets very close to its prey and ends up killing it. This literally refers to the damage of the artificial man. Human lays the foundations of his artificial world by keeping the space-time continuum and the truth in him to a minimum. In this way, it hurts both himself and those in the universe.

4.1.12 THE NAMES THAT REPRESENT THE OBJECTS BEFORE CREATION

nothing means not knowing

if humans start to know

that thing will be shaped in their mind

and will happen in the universe

How much did Allah teach the Prophet Adam about the knowledge of dark matter? Perhaps the correct question would be more like this: Which period of knowledge did Allah teach the Prophet Adam? Is knowledge finite? Either of these questions will provide an answer concerning temporal creation: Did Allah create things first or give their names? Should he gave their names first, which seems to be so, then He made His servant a partner in creating the good, the evil, the world and the hereafter. Therefore, every human being creates his own two worlds as a participant by the Allah's will to the extent of the words and the wisdom of the words. In short, things are created in a certain order and interaction. This is because the expressions and requirements of ignorance are in the stages of knowledge, and therefore, there must be compounds that base reality on that knowledge.

Since the duty of the objects and man is the word, there is a dark, invisible and visible substance of the universe and its creatures. They are interdependent in real and functional terms. Dark matter shows the effects of gravity so that it can be proven. The reason why dark matter is not visible is that it does not leave light or radiation. This is because only when one touches knowledge about an object does light spread from it emanating that knowledge. But isn't that normal? An item with no name or knowledge cannot emanate light. What's more, it keeps its light within itself as another form, that is, it waits to emanate it in the form of light. Due to this, today, the discovery of knowledge regarding dark matter plays a role in determining the speed of visible matter.

4.2 EINSTEIN'S *THEORY OF GENERAL RELATIVITY* AND FLORENSKY'S *REVERSE PERSPECTIVE*

4.2.1 DARK ENERGY

A similar situation occurs with dark energy. Dark energy helps us explain the expansion of the universe. The reason for this expansion is that dark energy and matter have scientific data such as their own physics and mathematics apart from symmetry and asymmetry. The reason is that the movement in the physics of unseen objects according to visible objects is in a waiting position, i.e., a passive state.

The string theory, which is based upon multiple worlds, was developed to explain the interaction between known and unknown substances and energy. This theory allows the connection between thin strings and different universes, and it measures the attraction between them. According to these measurements, there is a weak interaction between the universes. When knowledge becomes active through information, the elements of activity take their own forms. In other words, when the ratio of the active rate increases to a passive ratio, there is a strong interaction. This is related to perspective and reverse perspective.

4.2.2 HUMAN PERSPECTIVE AND THE FIRST HUMAN

And [mention, O Muhammad], when your Lord said to the angels, "Indeed, I will make upon the earth a successive authority." They said, "Will You place upon it one who causes corruption therein and sheds blood, while we declare Your praise and sanctify You?" Allah said, "Indeed, I know that which you do not know." (Al-Baqara 2:30)

The above verse starts with *Mention*. What does that mean? Was the Prophet Muhammad Mustafa present in the exalted assembly of angels gathered before Allah? But it is understood from the verse that Adam had not been created yet. However, if Muhammad had to remember something, was he himself the first created man and prophet? Why, then, did Allah hide the first man and prophet from the angels?

There are authorities in the assembly of angels because Allah commands Satan, his adversary, to leave his position. If we take the four great angels, whose authority is great, then where is the Prophet Muhammad's authority in this assembly? Why does Allah inform the angels that He will create a human being? Don't the angels yet have the power or understanding to carry the knowledge of the Prophet Muhammad?

The secret in Allah's creation of human beings is hidden in the fact that all names were taught to the Prophet Adam, while even the angels did not know as Allah asked these names to the angels. The angels didn't know. They praised Allah by having *admiration* and exonerating Allah from all deficiencies. Wisdom in the names or the knowledge of objects is understood from this dialogue between Allah and the angels. Allah underlines that he is aware of all things and shows that nothing can be hidden from Him.

Allah gradually created the order of creation. During this stage of creation, the last was the bodily existence of the human being. Like in the Surah Ikhlas, why did Allah, who is not in need of anything while every other being is in need of Him, create?

Based on these reasons, according to the knowledge which people have united around, Allah wants to be known. This is possible through knowledge, i.e., through the verses of Allah. These verses are the blessings He offers to people. Sometimes, people do not accept or ignore some information or create an artificial world for themselves. In such cases, being ungrateful and rebellious is inevitable. For this, it may result in some people being less aware of blessings, because whoever is not grateful will be distant from Allah.

Allah's curse is upon those who do not have faith in the blessings. This curse is an irreversible deviation. To deny what Allah has sent down leads to a differentiation of the heart. And this results in materialism. Materialism increases as less religion, faith and science one has. It is widely known that materialism has caused nothing but harm. Materialism means to live a life of dualism distant from the truth.

In another context, everyone has his own values of good and evil. However, the velocities, densities, and parameters for the composition of these dualities in regards of forgiveness, wisdom, and science, and their parameters, makes visible the fact that the true right and wrong change based on the knowledge. If such an approach remains unfounded in the sense of the consciousness of Allah and surrender to Him, it turns into the deception of self, without the beauty of cognizance. In a way, ambition and arrogance become dominant. *Have you not considered those who left their homes in many thousands, fearing death? Allah said to them, "Die"; then He restored them to life. And Allah is full of bounty to the people, but most of the people do not show gratitude. (Al-Baqara 2:243)* He gave the supreme abilities to the Prophet Muhammad. Therefore, by the permission of Allah, the prophet could see the time when something happened or would happen.

In spite of there being thousands of years between the places where he appeared and saw...

4.2.3 THE COLOR AND TONES OF REVERSE PERSPECTIVE

Before Allah created the seven heavens, the sky was nothing but a lot of smoke. (One day He will make everything return to that first stage.) Then Allah created the heavens and appointed every heaven with a task. One of the important points here is that the most important parameter is the proportional increase and decrease of the asymmetry from the seven heavens to the closest to and furthest from the world. This is because the structure and distance varies between truth and dichotomy according to this asymmetrical structure difference. To be subject to the intensity of the truth and to move it, one must reach two points, taking one as the beginning point and leaving that place. What is mentioned here is the Fatiha, and structurally it is an atom.

Consequently, being conveyed on the straight path by the Fatiha means reaching the truth through Allah. The other is the end point, the truth without losing its existence, the reverse perspective, is to live hand in hand with the truth as much as possible. Regarding the principle of integrity, one who grasps a particular thing and is enlightened regarding everything about it reaches to the other and applies it, and vice versa.

The Fatiha and its atomic structure are very deep building blocks that have been systematized within all the names and titles of Allah. Allah will show His verses to people in their own souls. When one adheres to His rope, it becomes clear to him that the Qur'an is the truth. Even those servants and angels talk with each other on earth too. This friendship on earth continues in heaven. Only those who are patient can be granted beauty. These are those who turn evil into good.

4.2.4 REVERSE PERSPECTIVE AND MULTIPLE WORLDS

In the Qur'an, there are examples in which the perspective of the multiple worlds are reversed in terms of temporal disintegration, i.e., the simultaneous, prior and subsequent features. From the perspective of concurrency, angels pray for the servants wishing for Allah to forgive the believers. In the previous case of the perspective, we may consider the words of Allah in His council with the angels. Allah commanded Satan to leave his authority, as he didn't prostrate before the Prophet Adam due to his arrogance. A similar case is with the Prophet Adam and the Prophet Eve's deception by Satan and being kicked out of heaven.

What happens after reverse perspective is the differentiation of the hereafter for the believers and hypocrites. A conversation takes place when Allah asks His prophets about their situation on earth and between those in the hereafter. In that, they ask each other about their situation on earth against the great beauty in heaven. One of the speakers talks about a friend of him on the earth, who doesn't believe in the resurrection. Just while trying to say that even he is in heaven, Allah calls those who talk. He asks them whether they know the truth about this person who does not believe in the resurrection. The one who is talking sees his friend in the middle of hell. Then he states that his friend has almost prepared his own end. The place of those who do not have faith in Allah and the hereafter is evident from today; and Allah is the owner of all wealth.

4.2.5 EINSTEIN'S *THEORY OF GENERAL RELATIVITY* AND *FLORENSKY'S REVERSE PERSPECTIVE*

kneeled down
towards the Kaaba facing him

the most beautiful thing

was to see everything from there

the beautiful prayer
was actually looking at himself

After the failure of five-century of this experience, it seems that we have no choice but to admit that the image of the world created with perspective is not a state of perception, but as a result of the demands of as strong and highly abstract ideas as possible. (Reverse Perspective)

Pavel Florensky used his characteristics of being a physicist, a mathematician and a pious person in his book *Reverse Perspective,* which became popular as his masterpiece. In his book, Florensky illuminates the paintings of the Renaissance period. From an artistic point of view, he says that the appeals and religious surrender in the works of that period meant to feel the expression as it is holistically, namely, the truth of the painting in all its aspects, and to feel the time of the work in an equivalent way to today, and to flow to that time, rather than examining the person who was looking at the painting and interpret it at that moment.

The quote above from Florensky's book sheds light on the issue that he persistently emphasizes. According to him, it is necessary to free the realm of the worlds of science and art. This is because a situation or concept can never be explained. In such cases, inappropriate meanings are imposed on it. In other words, the subject becomes a target of misconception in which the subject is removed from the reality in a centralist attitude.

Florensky's time period is also very important. As in all time periods, all the works done at the beginning of the 20th century come from the same ideas that support or are part of each other going in the same direction. A different example is the German philosopher of the 20th century Heidegger's understanding that takes the human from the subject and situates him as an object in the universe. Therefore, in the parts that complete the whole at that time period, while Florensky named the situation Reverse Perspective, Einstein, a scientist from the same period, called it *General Relativity*. Their common conclusion was quantum mechanics, though.

4.2.6 THE ASTRONAUT, REVERSE PERSPECTIVE AND RELATIVITY

Let's imagine an astronaut. This astronaut falls into *a black hole* on its event horizon. What change the astronaut undergoes due to the black hole's effect? In short, what is called the event horizon means that the perspective of relativity of the space-time continuum is the smallest or narrowest point, and this perspective corresponds to the deepest or greatest point of the reverse perspective. Thus, when the existing relativity becomes ineffective, the astronaut does not feel anything in a normal way. (In this sense, the

Prophet Muhammad's resilience against the situation in which he ascended to the miraj can be shown, or that Allah does not give people a burden that they cannot bear.) As there is relativity, the astronaut's falling at a specific point where the black hole is located in the context of relativity brings up the question of how this has happened. This is because, according to relativity, the universe has layers of space-place, which carry the traces of the astronaut or his effect. Besides, he may still be living in a completely different place and time mass.

With a sharper line, an object is reused elsewhere and this is called the usability principle. And in response to this principle, relativity dissolves and singularity already insists on resolution. This is because singularity is the resistance itself and represents the resistance of the truth as well. A circular expansion occurs here due tof the elliptical increase. This tells us that the astronaut plays a role in the unification of worldly knowledge and the depth of the dimension in his trait, as well as the dimensional information. But it is not limited to this only. Also, an astronaut is usedas material for other information and interim purposes. These materials are useful because of the information variability they have when determining the requirements of particles that can be used in all directions to create the resolution of information and content. In this sense, knowledge is all about how substances create them and how they are created through knowledge in return.

4.2.7 BELL AND NON-LOCAL INTERACTION

Local interaction is that the person or object is affected by what is done. For example, you and the lines you are reading now are in the same place. Positive or negative influences from these lines refer to local interaction. As with all things, local interaction has a counterpart to non-local interaction. The physicist Bell is the first theorist to point out the difference between the two subjects in his theories on quantum physics. What does that mean? On the basis of non-local interaction, this indicates that the place where the incident takes place and the affected place are different.

During his research, Bell focused on the physicist Pauli's work named *spin*, which is a popular concept in physics today. This proven invention of Pauli is the finding that there are two values in electrons. In fact, this is not surprising. Quite the opposite, matter reaches the culmination of its substance just like how its particles work. Just as planets have orbits, the smallest quantum particle also does. According to the space-time continuum, each trajectory is protected by another orbit that will absorbe it and it is supported to maintain its functionality. Returning back to Bell, while studying electrons in more detail, he noticed that they would rotate twice on their axis to return to their location. However, he realized that electrons carrying this duality were subject to non-local interaction only. In short, what Bell observed in his experiment is that physical effect reaches from one place to another without leaving a trace.

Negative examples regarding this issue are curse, magic or not giving one's blessing. In fact, the positive prayers that reach Allah prove how dangerous these concepts are. If it weren't so, those prayers wouldn't have had an ultimate target. On the other hand, what makes electrons so frenetic? Is it Raphael (PBUH) with the last *yellow* horn? But there is always an equivalent for everything. This is because either it is instantaneous or better ones, if all prayers are answered by Allah, evil thoughts and configurations will be of course accounted for as well. As we know, the Holy Qur'an tells us that even the earth will question us. Bell has proved this striking feature of electrons in atoms. This is in a way the teleportation of knowledge from one place to another. The question that should be asked again in the quantum dimension is whether the reason for the purpose in non-local interaction has the capacity to regulate time or not. The reason is that energy and matter depend on their purposes. Thus, they have directions in line with the gravity.

In other words, the knowledge in the substance or wave has a regular load of tolerance. Whenever the tolerance limit is exceeded or falls below the limit value and both cases become stable by repetition, a transformation or change occurs. An example of this in biology is that the molecules change the DNA by entering into the genes of a different cell. This change is both local and non-local. They result in enacting their own parameters. As in natural and unnatural processes, every law affects other laws and their elements. They impose their ability to respond to signals or a warning effect from their strengthened dimensions. Thus, in both cases, there is growth, and an artificial consciousness ignores the existence of the real consciousness covering the colors of truth. Therefore, this time, the light that emerges from the ambivalence falls below the limit. The atom or particle in the object or human being is imprisoned in darkness. This affects the universe and even the nature. However, nature will have its revenge.

4.2.8 NON-LOCAL QUANTUM

Continuity means that something affects or exposes many other things. There is no discontinuity in this context. The reason for this is that an impact is caused by many events, as well as by itself. This is natural knowledge. But the interesting thing is that an electrical thing can cause a magnetic field. How can something electrical produce a magnetic field? The reason is that particles, as the main center of substances, have the function of carrying electric current. These are cable-shaped material that can be transported without damage and without losing the speed (values) of the multi-distribution function. Of course, they carry this information with them. Otherwise, they wouldn't know when and how to act and react. In fact, just as there is electric current forming the magnetic field, the moment when the electric current is at a significant density, the simultaneous magnetic field is triggered as if responding to a call. Moreover, it knows that this response should be in the form of a magnetic field. The explanation is the presence of anti-electrons in the magnetic field itself. Only in this way, a situation occurs, such as a deviation, delay, or overlap.

Whether something is *anti* is determined by the task it carries. Not only the electricity produced by the electrons in the realization of the infinite crossing but also all the tiny particles of an atom or molecular compounds play a role in scaling them to their extent of effect. The cause and effect are from the same element. Despite spatial non-local situations, they complement each other partly in temporal non-local situations. For this reason, they can only carry the same purpose and ask other particles to establish a pattern for crossing the magnetic field as in this case.

4.2.9 CONSCIOUSNESS AND CROSSOVER

Crossover events are not finite because a crossover necessarily affects the entities in the universe as close to and far away from their concurrent dimensions, i.e., in visible and invisible form. The transmitters of interaction are also the antecedent of the article or element itself, which is the main center of the divalent system (e.g., matter and anti-matter). The reason why the crossover is infinite is that everything has a definite center. Although centralization cannot remain constant because of variability, or to accommodate it, the universe has both consciousness and a heart.

According to the consciousness of the universe or humans, the crossover system works in the same way. This means being able to live in a different situation or in another location without the need for a physical environment or to be connected to it. The knowledge of this lies under *how to get out of the space-time continuum* and having reached a certain potential. We can see these clues in the verses of the Surah Al-Kahf and in the verse below: *O company of jinn and mankind, if you are able to pass beyond the regions of the heavens and the earth, then pass. You will not pass except by authority [from Allah]. (Ar-Rahman 55:33)*

In today's quantum physics, the subject of teleportation is studied as a branch of physics. Now, what is the contribution of consciousness to the systematic study of the crossover? That is, the experiences (crossovers) form all lines of time, and the order of how they are created is distributed by consciousness singularly. There is no other essence other than information in the numerator and denominator of consciousness. For this purpose, when a synthesis of intuitive expressions of integrity is formed, it is obligatory to take only a particle from a single piece of information and interpret it according to the dimensions of awareness within an ambivalent system of vitality or self-esteem. There are variables of one's awareness according to his interests, skills and occupation. The result to be concluded here is that everyone tastes the same honey differently. In fact, this difference is a condition of the universe. This is because everyone writes a part of the history of the honey based on their own taste of it and the nature of their own (tasting) sense. Everything is written within both himself and the universe. In the scientific context, there are rules in which the crossover has interpolation (in a sense finite based on the strength of the effect, i.e., significant) and extrapolation (in a sense, providing opportunities for estimations according to the criteria, i.e., not yet clear for clarity of expression).

In other words, they are either hierarchical invariants or formed from them by influencing the variability of causes and effects that suit their potential. Consequently, a crossover does not mean that there are no other conditions. Furthermore, the crossover is double sided. First, it compensates for the imbalance of the compounds resulting from asymmetric mobility. For this purpose, it converts some particles into a stable mass. Second, as a result of the first task, the gravitational field provides a perfect amount of gravity in the quantum fields.

4.2.10 QUANTUM AND RADIATION

If Descartes's Cartesian philosophy, which indicates the intuitive and deductive aspect without separating physics and metaphysics from each other, can be applied to the field of the fullness of reality (nonlocal) that is subject to quantum mechanics, then the thought of transportability of crossover quantum situations does not fall into contradiction. Actually, progress has been made today at the least regarding the quantum teleportation.

Teleportation means that when a substance or energy is transferred from one place to another, it is located directly in the other place without passing through the physical space between them. It has been rumored that the Prophet Khidr was able to be simultaneously present at many places without going through physical space. So, is it just a rumor or a distinction of information? Thus, while the Prophet Khidr had the potential to exceed ambivalence through his knowledge of time, an ordinary person would be imprisoned in the multitude of places created by this dichotomy, and therefore multiply his destiny instead of leaving his space. Is this the reason why Allah emphasized in the holy Qur'an that the previous ones had obtained more knowledge? Perhaps with this, Allah was pointing to the fact that the ropes of the knowledge of truth have been curbed by the increase of duality. Since the knowledge of truth behind the scenes increases considerably during the end of life, there is also a need for science then as well. This is quantum mechanics.

4.2.11 ZEILINGER AND TELEPORTATION

cry uncontrollably
the day is over
today as well

beyond the sounds
that girl
her hair a bit black and a bit blond
a heavenly face

the age I desire
maybe ten
maybe eighteen
her hands on her back
standing as if she were forty

The Austrian theoretical physicist Anton Zeilinger seriously studied teleportation. According to him, the meaning of teleportation is that atomic particles (for now only measurable quantum objects, i.e., atoms and electrons; when knowledge and information about matter-wave transformations is sufficient, even substance will be able to be transported) can exceed the spatial distance by having effect without the need of time or physical transportation. Therefore, since teleportation is not a physical event, it can also be categorically related to metaphysics. What exactly does that mean? If we are about to teleport, why should that even be necessary? What systematic and scientific wonders, namely, truths do we lose so we need to be irradiated when the irradiation occurs? Is it a disaster, a light, or both that always represents the balance within dualism?

As a result, every situation or everything else is like an absolute medallion. There are two faces on a coin, heads and tails. So, what does that say? It shows that everything is a part of some other existence that complements it, and that it concurrently aligns and joins another situation at the same point. The distance between two things can be as far away as the distance between the seven layers of the sky. The particle holds what the whole also includes. This totality can be found in the particle. If we assume that the same things are valid for human beings in terms of distance, is it wrong to think that people are living outside the solar system?

There is another very important point that comes to mind. In the context of relativity and reverse perspective, which lives are we currently living in the solar system that are mentioned in the Qur'an? On what planet does Moses live or will live, and when will he fight against Pharaoh? Humans experience death, but will everyone actually die on a single day when the horn is blown?

4.3 ALLAH CAN ABOLISH ANY VERSE HE DESIRES

4.3.1 IS IT IRREVERSIBLE OR AN ECOLOGICAL FOOTPRINT?

Stephen Hawking, an expert on black holes, changed his mind regarding his long-term theme on black holes on which he had worked and insisted on. In that, he became convinced as a result of his scientific debates with physicist Leonard Susskind. Moreover, he said, *There is no loss of infinite information as in the black hole, but the loss of knowledge takes place when something becomes different or goes into different shape.* In the case of Hawking's irreversible loss of knowledge, for example, cold water is poured into hot tea, here the information gets lost with regard to where the warm-fast water molecules and slow-cold ones are. The current situation corresponds to the increase in entropy of the total system and the loss of this information. And this happens through irreversible processes such as the principle of classical thermodynamics. The incident, which refers to these processes, is that in a closed system entropy can only increase or at best remain the same. But, Hawking here adopted the rules of classical physics and never entered into the complex form of quantum mechanics. Quantum mechanics is changing the infrastructure of the work by addressing the inter-communal systemic areas of living. This is because there is no loss of knowledge in the context of the ecological footprint. The ecological footprint can be considered pros and cons within the man himself. One can be generous on the one hand and extravagant on the other hand. On a large scale, this refers to inter-country rises and falls. In a developed country, the number of people per square meter is greater, while it is less or proportionally less in less developed countries. To sum this up, if there is a lot of consumption in one place, consumption in other places is definitely reduced. Thus, the values of the cosmos are always the same and remain so. This is already important for the system of inertia. Thus, this system presents the reference points at the level of its symmetry constant for the solution of dual-quantum problems in the form of polarization as the dimensional rich and infinite network for geometry and topology. This example also shows that there is an elliptical increase in the shape of a circular expansion.

In other words, the knowledge of the world and a person's character play a role in his worldly knowledge, property, size, depth and connection of dimensional knowledge. Another characteristic of one's knowledge of the world is that it presents itself as material for further knowledge and intermediate purposes. These are useful in the need for each versatile usable particle as a solution and content for the potential formation of knowledge. And in this sense, information is all about how the material forms it and how the information is designed accordingly. In short, there is no loss of knowledge as per the principle of context. On the contrary, there is congruence to the proportional state (gravity), and infinite causes are provided for other dimensions of simultaneous entropic possibilities. Entropic possibilities are associated with infinite possibilities of variations in the materials and dimensions of knowledge.

4.3.2 IS IT RECYCLING OR BLACK HOLE?

What happens if you build palaces?
are they without beams?
what about your lives without beams?
have they ever existed?

have you ever met a hero?

have you thought they were invisible?
when he was ashamed of his morsel
did you see the look in his eyes?

have you ever been quieter than a snowflake?
have you ever tried it?
could you have looked at the horizon
without abandoning your roots like a cypress?

Man builds and that becomes his home. Then, he lives in and sells the house he has built. Now, the buyer lives there. And after many years, the house is now old. Sitting at that home doesn't appeal to anyone's dream. Then, since no single person is interested in it or due to landscaping causes, it is destroyed. A brand-new splendid house is built in its place. The foundation of the house is never the same. Even though the shape changes, the basic materials remain the same. What should not be forgotten is that the material has been transformed into a post-destruction phase (recycling). This means nothing disappears; it just changes. But when will there be houses without the need for change through destruction or modification? The answer to this question is only found in houses that have no pillars or beams and that are resistant to years just like the creation of the universe. However, human beings are not capable of this (but Allah makes it possible too), because their structure should not respond and give direction to change. After all, this is not the meaning of the revelation flowing in the universe. Its meaning is that man behaves in accordance with his ability to characterize the activation of his nature.

4.3.3 ALLAH CAN ABOLISH ANY VERSE HE DESIRES

According to Stephen Hawking, if there is any knowledge in radiation, then although there are particles or photons, they must be aware of one another as a single quantum of radiation. In other words, he was of the opinion that they can never be independent from each other. When these partner particles show the same response to a certain amount of knowledge, they perform a proportional crossover to the mass by affecting it together. The remaining ephemeral remains of their existence passively remain sprinkled throughout the field due to insufficient knowledge. Since each radiation particle, which has escaped from its own orbit or mass, is already in the stage of being crossed over with its virtual partner, futures may be created through the intensity of this infinite cause-effect. Therefore, the loss of a unit of information in a black hole means that the future of the unit related to that particle is no longer obvious. Whenever the order does not find the specific particle it is looking for, it sends out a different identity search signal. Whichever unit responds to this call, in particular, to these signals, then that unit prepares the identity data that is missing but necessary for this balance. The content of these signals can be readily responsive to changes on their axes, rotating with each other and showing small deviations. The order in nature is perfect. But Allah's will is another thing. In other words, the verses revealed by Allah are flawless in every aspect of function and structure. Again, as described in many verses, such as *We do not abrogate a verse or cause it to be forgotten except that We bring forth [one] better than it or similar to it. Do you not know that Allah is over all things competent? (Al-Baqara 2:106)*, the potential power of Allah is highlighted. But do these verses tell you that some of the information falls into what is called a black hole? Or does it mean the variability that is fed by the movement and the mobility of nature, which is realized by the crossover? There is no loss of knowledge. Rather, there is compliance (gravitation) according to the appropriate situation. At the same time, infinite causes are provided for the dimensions of other entropic possibilities. The possibilities are related to the infinite ones in the variability of the material and size of knowledge. And so, there is a correlation between chemical processes.

4.3.4 GRAVITATION AND QUANTUM

The theoretical physicist, Stephen Hawking, recently changed his previous thoughts regarding the black hole that swallows all materials and radiation, around it. The physicist, Leonard Susskind gave a detailed account of his own opinion against Hawking's and his argument with him regarding black holes that lasted for three centuries in his book *My Battle with Stephen Hawking* written in 2008. According to Pushkin, the black hole and information do not disappear forever. In the book *The Reareno Black Holes*, he points out that he now looks at his thoughts from a different perspective, as indicated by the title. Also, since the 20th century (Maldacena managed to do so partly), like all physicists, Hawking also had a dream, which was to explain classical physics and quantum physics together. The fact that classical physics is only a part of quantum physics proportionally shows how much the atom is understandable. And this is obviously not too much. As a result, what people know indicates the 4% of their knowledge regarding the atom.

Although there is not enough knowledge to ensure integrity in gravitation and quantum theory, the situation that was proved in the '70s is that the quantum truth and relativity affect each other. Perhaps, the black hole is not a huge and alluring object. As it has been believed, the black hole does not have an event horizon where light and matter cannot escape. The assumption of the absence of event horizons makes the normal functioning of gravitation and quantum mechanics common. When the black hole event is perceived as a necessary collapse, the inward collapse or collapses for the day and night, there are transitions in the points of intersection between light and dark. In this case, physicists are one-step closer to find the formula for the world.

4.3.5 THE EVENT HORIZON

To explain the event horizon through human being who is a part of the universe provides a different dimension. The most common reason for this is that the universe or human being is not static. Both truth and dualism are moving in certain dimensions. This is because they are the monopoly of the soul that provides both movement (truth) and mobility (dualism). On the other hand, the existence of the self creates the illusion of an event horizon. This is because during rapid events, people tend to observe the loss of their appearance. Even though events are a whole, their conceptual perception according to human nature is only possible if they have reached a certain potential. External events make themselves feel in the end in order to be thought upon. The effect on the body by the conclusion reached in combining the experiences explaining the event horizon with the experience of not having an event horizon can be shown as an example. In short, scientific explanations according to knowledge and its nature may lose their meaning. And from the religious point of view, all knowledge reaches to Allah, *And when We substitute a verse in place of a verse - and Allah is most knowing of what He sends down - they say, "You, [O Muhammad], are but an inventor [of lies]." But most of them do not know. (An-Nahl 16:101)* and some verses lose their judgment while others pass it by. But what needs to be clarified is the state of the verse that has lost its verdict. Does it disappear or is a level of knowledge surpassed by something else? If recycling that is an indispensable parameter of creation is taken into consideration, it is obvious that the latter option is more accurate.

4.3.6 BLACK HOLE AND CREATING

If He wills, He can do away with you, O people, and bring others [in your place]. And ever is Allah competent to do that. (An-Nisa' 4:133)

A black hole of course swallows some data. However, it makes interim creations according to the needs of differing units. According to Allah, the creation of all human beings is as easy as the creation of a single one. For this reason, there is a localization that should be asked when the above verses are considered. Where is the place that Allah calls the abyss? This could be hell or another place. But let's say hell isn't mentioned here, even it must be better than this nonexistence. This is because it means that Allah does not punish them even with hell.

4.3.7 QUANTUM, GRAVITY AND IRON IN SPACE-TIME

The phenomenon of gravity related to classical physics and quantum theory is different. The formation of the former is more regular. The first one relates to the formula $E = mc^2$ of Einstein. However, it is not possible to place the second one in such a general equivalent. This is because in the quantum, gravity is formed according to the conditions created by the infinite variables of the atom or it is necessary to calculate a very high level of mathematics because the solutions are connected to the extension of gravity. In other words, to predict each quantum gravity measurement, it is necessary to know all the findings, i.e., the particle's formula of the world. It can be possible to explain the speed, distance, transformation of the matter into a wave, and vice versa by removing time, in a sense, by eliminating the concept of relativity in the timeline. This is because quantum gravity expresses itself not by uniform particles but by ones that complement each other, i.e., the aim is inductive, and the particles must be infinitely variable so that existing shapes can respond to their needs. This means, in terms of quantity, that there is a proportional distribution of variability in the chemical states of the fields of gravity.

What makes gravity both a constant and a variable? Particularly with the involvement of internal and external factors that have dimensions of inertia and duality, these dimensions make sense of the presence of asymmetry on the one hand and symmetry on the other hand. The mass of symmetry derives its energy source from a very different quality. This duty is undertaken by quarks instead of electrons. The role of iron here is great in that it enters deeply into the atomic nucleus and gives power to the quarks by transforming them to undertake the role of electrons. Thus, the atom, the nuclear membrane and the nucleus become subject to another form of encryption. In fact, the atom reestablishes itself in its internal structure seven times or in seven dimensions and takes another identifiable value each time. The success of the iron element during each period exceeds space-time.

4.3.8 THE WAY THE ATOM IS DIRECTED TO MECCA

So from wherever you go out [for prayer, O Muhammad] turn your face toward al- Masjid al-Haram, and indeed, it is the truth from your Lord. And Allah is not unaware of what you do. (Al-Baqara 2:149)

These verses are the commandments of Allah. A word or a sentence is more or less applicable to each field or situation. Considering this, it is understood that the work, the body and the soul turn towards Mecca or are within it themselves. In all that he does, in all things and prayers, the person who turns towards Mecca or knows that all places are qibla, in every circumstance, he takes Allah as a witness and purifies himself from his own evil. Where does this purity lead the servant, and what is bestowed upon him here? If the purification of the servant, in proportion to his mistakes, comes to the particular state that Allah demands, it will turn out as if the person did not make that particular mistake. The traces of this previous negative attitude and behavior are wiped out.

Allah abolishes any verse or any provision in that verse whenever He desires. As in the case of the black hole, the evil that a person does is transformed into goodness. But if a nation does not deserve enlightenment, the light of Allah is withdrawn from the environment. With such a withdrawal of light that serves as a means to reveal beauty, there remains the incomprehensible expression of duality. That place and the people in it get drown in the darkness and evil in an unstable state. As the Qur'an says, *They were killed by the equivalent of the evil they created.*

A different dimension of this is the situation of being dislodged in a place due to the reflection of the light and/or its manifestation. The reason is that the structure of the earth and the sky can draw a picture, which cannot remove this light. A single heart can withstand the presence of the light. Only that single neutron reduces or prevents the development of decay as much as possible, for Allah has filled the heart with love. Only with this love in his heart, which is just as sharp as a knife according to the human structure and the potential for mercy that are never dissolved but solute, it is possible to soften the weight of Allah's blow from His own soul. In a way, the working of iron is possible with the sensory and organs of the servant's heart. While man can flow in the orbit of divine light with such light in his heart and even become the divine light itself, why does he still worship his fancies and desires?

Two events from the Qur'an can be shown as the miracle of the heart. *And when Moses arrived at Our appointed time and his Lord spoke to him, he said, "My Lord, show me [Yourself] that I may look at You." [Allah] said, "You will not see Me, but look at the mountain; if it should remain in place, then you will see Me." But when his Lord appeared to the mountain, He rendered it level, and Moses fell unconscious. And when he awoke, he said, "Exalted are You! I have repented to You, and I am the first of the believers." (Al-A'raf 7:143)* and *But Allah would not punish them while you, [O Muhammad], are among them, and Allah would not punish them while they seek forgiveness. (Al-Anfal 8:33)*

4.3.9 EMOTIONS, COLORS AND SPACE-TIME

The Sun, Moon and all objects are created according to certain measurements. Allah has given the objects their disposition. The features and the meaning of existence of the Sun are the same. The Sun heats the Earth and illuminates it. The scene of color is noticeable in light. The earth and its crops need the Sun and its effects. Therefore, in the cosmos, each object acts within its own orbit. Each object is connected to other ones according to their task, and the color and size of them can be measured according to this principle of commitment.

4.3.10 THE CHANGING OF A BLESSING AND QUANTUM CROSSOVER

If Allah were to goodness within the disbelievers, He would make them hear what He says. But even if He makes them so, they would not think on what is said and would turn away. The influence on them wouldn't last long. However, by following Allah's commandments and fulfilling them, the servant would prevent the formation and domination of such mischief.

Obeying Allah and His prophet upon his call to faith is to know that faith is life itself. When one believes in Allah, His demands and His creations, in short, the moment he mentions the creator, He is now in the heart of that person. When Allah manifests in the heart, all the particles in the body come together. However, it is not just the body that is gathered. According to Allah's manifestations in the universe, objects also move in His orbit. For example, the servant obeys Allah based on his knowledge by living within compositions of two dimensions and the weight of certain substances.

What appears of Allah in the servant's heart is merely as much as this pure state within this way of obedience. The bodies of the servants and objects in the universe quickly respond to this purity. They become the propeller with the parts that drive themselves around.

Similar atomic and subatomic situations are also called into question for bad attitudes and behaviors. It is important to avoid negative factors, for evil not only reaches the ruthless ones, but also spreads all over others. As it was told in the story of the Prophet Adam's sons, Cain and Abel, Cain, who was jealous of his brother, the obedient Abel, killed his sibling and thus undertook the sin of Abel, who had been dead for a time, as he said. Even those who have betrayed Allah and His prophet have betrayed their own religions. The traitors are those who do not believe, but it is a call to the believers to remember and relate their religion. Moreover, if there are prophets among those who pursue evil, Allah will not bring down punishment upon that community. In addition, Allah does not punish them as long as they ask for forgiveness. And for those who believe, if they fear Him, Allah will give them the ability to separate good from evil and thus lead them to victory.

In another sense, Allah, who knows the essence of hearts, puts fear in the heart of the enemy while instilling in the hearts of faithful people confidence if He wants to bring them to victory. According to a particular expression in the Qur'an, Allah's words take place with a reduced representation of the number of the two sides. The difference is that while believers feel a sense of victory, the enemy has a feeling of arrogance as he thinks he will win. This leads him to defeat. This is because if Allah wants a task to be fulfilled, He determines that work to take place at a certain time-space. In a similar vein, if one hopes and believes something in a healthy way, and each time gets closer to it step by step, then Almighty Allah will complete his task to the end.

Every person is created with certain features. Therefore, a person's desires for and belief in a particular thing occur with the particles according to the belief in that thing. Magnetic particles accumulate one by one. Then, the work desired and believed in the magnetic field intersects. The point to note is that all things created by Allah in the universe are a blessing. His prayers and greetings are all blessings. The first of this, as well as the nourishment of the soul, are the book and wisdom itself. As given to the prophets, these blessings are given to all people to create their own balance within themselves, and then to keep this balance there with perseverance through carrying it into different dimensions with the truth. However, because the human being is not happy in the world or in his worlds, he wants other things. Thus, Allah wants to replace the blessing He has given. But for this, He demands that the person put into action his good-evil perception. This is because the pluralistic system of truth must be mobilized in one's will.

4.3.11 STRENGTHENING THE BELIEVERS WITH THE ANGELS

Another good news that Allah has given to the believers is that the angels will support them. The number of angels increases according to the mentality of patience in the servant. Just as human beings cannot count though they want to count the blessings of Allah, His patience is eternal and adjoining. However, patience and perseverance are possible only through the remembrance of Allah, who says that the success of patient servants comes from themselves. Nevertheless, whatever is going to happen will happen. The fact that there is any other information higher than this reminds us of a system or a systematic structure and other system(s) higher than these.

There are systems that are hierarchically reported to people by Allah that can vary over time. There are also unchanging systems encompassing these ones. The knowledge of all of them is in the book next to Allah, i.e., the al-Lauh al-Mahfuz. Man may not be aware of this knowledge but experiences it. The enormity of this horizon leaves human speechless. Sometimes insignificant things are conducive in the collaboration of people. Allah said to the Prophet, *And brought together their hearts. If you had spent all that is in the earth, you could not have brought their hearts together; but Allah brought them together. Indeed, He is Exalted in Might and Wise. (Al-Anfal 8:63).* What Allah says happens.

4.3.12 RUMI-SHAMS: THE WINE OR THE ROSE?

Then Adam received from his Lord [some] words, and He accepted his repentance. Indeed, it is He who is the Accepting of repentance, the Merciful. (Al-Baqara 2:37)

they said
this plate is yours
flawless white
fill it with repentance

I have the plate in my hand
one left and one right
but what kind of repentance
would fill it?

after many days
said the shadow of mine
that love is the plate
already accepting of repentance

One of the most fundamental assumptions for human to be the truth itself and/or reach the truth is repentance. The keys to opening the door of faith are given in the Qur'an. One of these keys is repentance. The one who repents travels towards Allah, who makes those on the way towards Him feel the eternal gates that can be opened through repentance. The servant learns to continue on this path. His repentances tell him everything. What is repentance? It is a purpose as well as a means to melting sin just like snow. The best word to explain how repentance is a purpose is that it brings an instant response of repentance as much as a servant's repentance. In other words, it is the touch of the servant to the word of Allah by coming down to the dual world and their worlds and lifting the veils. Every touch is purification. Every return from truth to duality is a verbal or nonverbal, visible or invisible announcement, which is brought from there to everyone and everything that is created in duality. It is the service of the repentant human to move from the created to beauty from divine light to destiny. And it is the closeness of the help and infinite compassion of Allah, because hope lives within repentance, and repentance to Allah turns the sin committed into goodness. Moreover, every committed sin can be repented for.

If a sin is committed and remains without repentance, Allah punishes the servant with more misery. Man is not able to have enough knowledge of space-time to calculate his misalignment. This means that he cannot know when, how or how much his punishment will be. But if the servant knows his servitude and repentance, he will say in his prayer, *"Allah! Help me! Forgive me! I live for you, and it is my duty to die for you."* One who has not buried his heart in evil knows that only Allah will protect him from things that He is not pleased with. This is because Allah never leaves the one who believes in Him alone. He transforms his evil into good or helps His servant and declares his position to the entire universe as elevated.

Here is a well-known story: Shams asks Rumi to go to a tavern and bring a bottle of wine. This is actually a test of surrender. Rumi does what he is asked and brings the wine. Shams takes the bottle that Rumi has brought and flips it. However, what comes out of the bottle is not wine but rose water, as the wine has turned into rose water.

The lesson here is that sin and good deeds are very close to each other. The phenomena that are perceived as miracles are bestowed, or more precisely revealed, by the Almighty to give wisdom to someone who has surrendered to His consent and love. When the situation is taken into account in scientific terms, the fundamental feature of all transformations is that meaning is subject to the universe of knowledge. This means that the invisible appears in the visible. In this regard, quantum mechanics exemplifies the fact that both possibilities are based on their probabilities and all materials have the ability to transform into any material desired in themselves. The matter that both expresses this phenomenon and realizes it is the possibility to turn copper into gold and the fact that all the essence of matter that makes this transformation possible has the same purpose, which is to serve Allah by His appreciation.

4.3.13 THE FUNDAMENTALS OF LIGHT, WAVE AND LASER

In 1905, Einstein's theory of relativity echoed worldwide. But Einstein received the Nobel Prize for physics that year, as he was able to explain the nature of light as both a particle and a wave using the dualism method. According to him and other theoretical physicists, there are coincidences and possibilities on the basis of quantum physics. But it's not just that! The most interesting aspect is that light is no longer a subject to be observed only.

It is no longer the only matter examined in quantum physics. Physicists play a significant role in this. This is because the way a physicist sets his ideas on light in his experiments also influences the characteristics of that light.

Therefore, quantum mechanics contradicts Einstein's religious beliefs and his very famous saying *Allah does not play dice* defines the situation.

Although Einstein endeavored to refute his own discovery of light and quantum mechanics to the end of his life, quantum physics had always proven itself in experiments. Yet Einstein's theory of relativity also questions the validity of classical physics. Consequently, quantum physics has taken place besides classical physics. If a physicist or a scholar can influence a physical experiment, then it is possible to say that whatever body is attributed to it, a form or non-formal state, and whether or not it is supported by this creativity, then the thing turns into the aforementioned attribution. According to the reaction parameters of physics, the physicist's mind and heart must have a characteristic of the object so that it can emerge. It should not be forgotten here that Allah is always in a state of creation with *Be!* and the rule of existence. Otherwise, there can be no explanation of this, and we are inviting the repetition of history by putting humanity and sanity in the center and accepting that the Western world is still in the enlightenment period and still influencing all scientific disciplines to this day.

4.3.14 EINSTEIN: *ALLAH DOES NOT PLAY DICE*

Quantum physics is considered as truth that occurs according to possibilities rather than being deterministic like classical physics. Determinism is the idea that the entirety of events taking place in the universe is predetermined by scientific, proven laws. Determinism does not just relate to physics, but it also has a structural function in the theological sense that everything is predetermined by Allah.

Diversity, words and their meanings are increasingly being used in different fields such as history, mathematics, and psychology, but with similar or different functions. If the doctrine of classical physics determines the laws of nature according to the concept of determinism, then it is useful not to consider this concept as disconnected from its religious meaning. Moreover, Einstein, in his article on *The Electrodynamics of Moving Objects,* published in the journal *Annalen der Physik,* explains through the theory of special relativity that space and time varies depending on the observer. There is no such thing as absolute time; everyone shapes it particular to himself. In this sense, time is relative. On the other hand, intuition is absolute because they are a whole. All names of objects are included in this entity. Human beings can only make synthesis according to their particularity or knowledge. The synthesis they make is again relative. Einstein did not receive the Nobel Prize for the special theory of relativity. This award was given to him for his work, which was based on the invention of laser. The scientific community, even Einstein himself, could not see the completion of the studies on the properties of light as the subject of the theory of relativity and quantum mechanics.

The name that comes to mind here is the poet and doctor Gottfried Benn. According to him, a poet always writes poems, which are similar and complementary to each other, and actually, he wrote only six poems for a lifetime. This is about all people rather than concerning merely poets. This is because a person's action is in line with his own characteristic. It is based on the worlds he builds and knows. Einstein also claimed the same thing. Therefore, we cannot say that Einstein's theory is incompatible with his other theories, because they are not and cannot be independent. If so, would the universe have had its own history and system up until today based on this history?

Albert Einstein could not adopt quantum mechanics. He resisted this theory for as long as he lived and put forth great effort to refute it. His aphorism *Allah does not play dice* explains why he opposed quantum mechanics. A theory that is founded on possibilities rather than truth and seems to be impossible to be refuted experimentally goes against Einstein's conception of Allah and therefore his own belief. But quantum physics is based on possibilities and attains the truth. It is only for this reason that it is evident. So the truth is within possibilities. Otherwise, odds or repetition could not have ever happened. Indeed, if only the truth were present, would the man's dualistic character of creation have any meaning?

According to Einstein, there is no mathematics that man can define. Thus, the truth in the form of classical physics and mathematics is called into question. As such, does quantum mechanics or classical physics overlap more with determinism? In summary, *Allah does not play dice*, but rather, as He created people asymmetrically according to dualism, *humans play dice*. What Einstein failed to take into account was this very matter. He did not consider the fact that Allah created us as human beings or the wisdom of reasons for dualism within His creation.

4.4 THE COMMAND OF ALLAH

4.4.1 CLASSICAL PHYSICS AND QUANTUM PHYSICS

All branches of science have their own specific formations and, according to Born, quantum mechanics cannot be formatted as per certain scientific laws, such as classical physics. Otherwise, quantum mechanics, whose accuracy has been accepted so far, should see its formalism as defective, and inventions, which concern this theory, should be refuted through experimentation. These two situations bring to mind one question, which is also the question of what scientific determinants are. The two parameters of the universe clarify this question. The structural function and functionalized structure of the universe, being interdependent and deterministic of each other, are both moving and variable. As such, classical physics corresponds to a characteristic that can explain only certain truths, while quantum physics has an asymmetrical structure based on these two parameters and carries its functionality from here to symmetry. Classical physics, however, ignores the peculiarities of creativity when considering these different situations as inexplicable.

A cherry tree depicts this topic well. The cherry tree has three dimensions. It becomes a sapling from its seed, a trunk from the sapling and finally it becomes an actual tree. In April, the tree blossoms and gives forth fresh, green leaves. In the last week of June, cherries ripen. This is the development of a tree that reveals more or less classical physics. What about quantum physics, however? The size of the sapling is never the same as that of another, and no cherry is the same as the other when it's in season. This is nothing more than following the rules of Allah and being subject to an asymmetrical structure and its functionality. In terms of the micro cosmos, since human beings depend on a dualistic way of thinking, it is imperative that there are infinite possibilities for the systematization of its inner and outer world with its dualistic forms. For this reason, it would not be possible for the elements to cause changes in these complex components for the variable and moving balance provided by gravity. The main factor in the formation of these changes is the light that the elements emit when they unite. Without this light, waves would not exist from the formation of these elements and/or elemental components from their length to the spectra of the elements. Neither classical physics nor quantum physics can be refuted, as said in the words of Allah: *We raise in degrees whom We will, but over every possessor of knowledge is one [more] knowing. (Yusuf 12:76)* As a result, thanks to quantum mechanics, not only what happened in the macro-micro culture based on all branches of science can be described, but also specific application areas in all branches of science can be found. For example, in medicine, the X-ray or organ transplantation requires quantum mechanics.

4.4.2 IS A THEORY SUBSTANCE ACCORDING TO THE UNDERSTANDING OF QUANTUM MECHANICS?

Whatever the substance is, it is not made of substance.

Hans-Peter Dürr

The parrot can do these things, because it has the ability to speak and fly. If a person is emotional, he perceives reactions and life itself. The same is true for theories. Theories have an essence, and the theory we are talking about here is quantum mechanics, which must have the ability to be substance and photons so that it can be valid in the fields of application. So, quantum mechanics is what it represents as a theory. This is because a dualistic behavior becomes functional within a dualistic system. What about people? Is man just living in a dualistic world or is he the one who has created the dualistic worlds as part of the world he lives in? Let's say so; then, he has the ability to do so everything that is visible or invisible, good or bad to himself.

4.4.3 WAVE-SUBSTANCE AND SUBSTANCE-WAVE

And if whatever trees upon the earth were pens and the sea [was ink], replenished thereafter by seven [more] seas, the words of Allah would not be exhausted. Indeed, Allah is Exalted in Might and Wise. (Luqman 31:27)

The theoretical physicist Born writes about why he tended to hold a particular philosophical stance in his scientific articles. He specifically examined the connection between truth and symbols. The reason why this subject gained some importance in physics is that the atom and light are both depicted in a fluctuating and fragmented form. In more detail, photons exhibit the characteristics of waves, and electrons behave like waves when they come in contact with light. This is proof that the basic nature of the universe is dualistic. This also means that the massless photon and the electron can be both wave and substance at the same time. How can this contradictory situation be comprehended or rather be explained? The problem is solved by perceiving both as symbols. Born saw this and used it as a tool aiming to reach the truth of these symbols. All objects have names and/or aspects that are known and yet unknown. Born thinks that all of this knowledge will be achieved through philosophy. But the main issue is that creativity comes before philosophy. Truth is only revealed through the creative combination of knowledge and is spread up to the seven heavens. This is the spectrum according to the density in which they are spread. For this purpose, it has been observed that the same, similar and/or complementary idea or system prevails in different parts of the world during certain periods.

Creativity is a situation that concerns not only philosophy but also all areas, especially art. In the Qur'an, Allah repeated a story from different aspects based on such situations that warranted it. The words are never used the same. The Creator uses His creativity through different combinations according to this flow. Creativity is the angels delivering knowledge through revelation or man pulling himself out and looking at knowledge or symbols resembling the commands of Allah. Both cases require a certain amount of knowledge about Allah. Knowledge creates irregularity. If there were irregularity in a person, there would be so many points creating the entropy that is so frequent and different. The presence of these points increases the likelihood that the spectrum will fall on them.

In other words, a person reads similar things and more or less has an idea about what he reads. The issues that he pays attention to according to his point of view and perspective also deepen his knowledge on that subject. As the person unearths the knowledge within that particular object, the knowledge about this object reveals itself quickly. If the size of the heart is not overlooked, the point where all knowledge intersects is not a one-way representation of the mind but rather an aspect of the heart. The right path with these truths is pure love.

Everything is a sign, atom and light. How about Fatiha? Isn't it a light, atom and symbol? Is the constellation of the Fatiha the same as that of the atom and its nucleus? Both create infinite variable signs. Both demand absolute truth from the distant and near crossover, which creates the space-time coordinates of the bivalent system. Crossovers are nothing more than previous gravity and are the basis of the next one as a way of preserving their mobility and movement. The universe is always in motion. Each time, a crossover takes place because Allah's command never falls short. The spectra of the light waves cross over according to gravitational volume and are instantly transformed into matter. The function of gravity is to form particles of gravity from itself for subsequent crossovers. To this end, it remains functional as matter solely, and thus an order is formed through linear gravity. This order represents a part of the truth.

4.4.4 THE PHYSICIST, LOUIS DE BROGLIE

According to the physicist Louis de Broglie, if the wave can be transformed into a substance, the substance must also be transformable to a wave. The experiments showed the accuracy of Broglie's intellectual work on matters and waves. In terms of their measurability, the masses of electrons are known to transform their substance into waves according to reverse transformation. Broglie concentrates on the need to destroy mass by assuming that the wave properties of these masses will show the same behavior as the waves show as well. This is because if the substance can turn into waves, then it will have the properties of a wave. Just as light is in waves and light stacks itself on top of each other (interference) to create the darkness, substance plus substance must be destroyed. But when the substance and its particles are transformed into waves, do they remain the same in terms of the volume and properties of the mass? Or does the substance, which becomes a waveform, take other features such as pliability when changing its form? Such questions have not yet been answered. In this context, how much does classical physics correspond to physics? In classical physics, the form of a system in which the known is intense is made by addressing the mind. What about the heart? Since the functionality of the heart is not affected by the continuity of the spectrum and it is not necessary in its formation, it accepts this feature as incompatible, which shows this as a reaction to emotions or rather unresponsiveness. This is because in the geometric shapes of the mind, the characteristics of emotions are almost invisible and they push emotions out of life to maintain and preserve order in the mind. On the other hand, an increase occurs in the darkness of the senses, which are lower than feelings. They direct the next spectra as a criterion and therefore develop a uniform system. They also absorb or imprison the geometric shapes that it desires. Thus, its mass remains stable making its growth necessary. This structuring can be systematically digested in all areas of life, as well as in humans.

From here, a heartless, and therefore, an artificial world filled with violence emerges.

4.4.5 THE TRANSFORMATION OF LIGHT INTO DARKNESS IN THE CONSCIOUSNESS

...And if it were not for Allah checking [some] people by means of others, the earth would have been corrupted, but Allah is full of bounty to the worlds. (Al-Baqara 2:251) The result of *light plus light* is darkness. But with the accumulation of the light, transformations to other stages occur. This, in a sense, indicates how night and day, weather, the climate and seasons produce changes. But there are a variety of stages, which are all important. They also have other tasks to the same proportion and perform them simultaneously. With them, they offer the necessary integrity for gravity so that the balance is redressed.

In other words, the weather report does not depend solely on geographical lines. At the same time, as a reflection of the behaviors underlying dualism and various events, the stages have many functions within themselves. One of them is the situation of avoiding evil through malignance, as mentioned in the Qur'an.

No matter what day it is, a child born that day possesses the knowledge of it as well as all the past and future. The way to reach all past knowledge of living man is by reading the universe and/or the signs within it. In this way, it is observed that every day writes its own history with the existence of the atom even in all stages, so that those born in the future are prevented from falling into the gaps thanks to this knowledge obtained. History basically writes its own timeline for the hereafter. Another area related to light and darkness is the phase of consciousness. In the psychological context, it is consciousness that is amenable to structuring and obsessed with a monopoly on negativity in the inner world of the person.

One can show how he delivers his consciousness up to artificiality and how light is destroyed. The transformation of light into darkness means that the festivity of color is lost. Therefore, it is the fear and all its kinds that the structures, which shift into these dark phases, unite and raise up. In fact, it is considered as a natural right to have defense mechanisms. It is important to note that there is specifically positive darkness and negative light as well.

4.4.6 THE UNITY OF THE WAVE AND THE SUBSTANCE

Although the cause has not yet been figured out, the wave can be transformed into a substance, and vice versa, as known. There is something else that has not been solved either, which is the subject of how the transformations between wave and matter are achieved and the result of their harmonious essence. The point where matter and waves meet is that with the organization of both by the soul, it becomes possible to merge them, and the necessary factors come back to soul as a combination of the two. Those who do not rotate are dispersed to be the ground for reunification in the body and the universe. Hence, the soul becomes the complexity of meaningful relationships. Without the soul, there is no opportunity to evade evil with malignance, and the universe becomes finite through its universal turnings. Although there are differences in size, not only human beings but also all things in the universe have hearts, minds and souls. Even all abstract and concrete concepts have them.

As a result, substances and waves in the human and the universe are transformed into souls through transformation. *But if they endeavor to make you associate with Me that of which you have no knowledge, do not obey them but accompany them in [this] world with appropriate kindness and follow the way of those who turn back to Me [in repentance]. Then to Me will be your return, and I will inform you about what you used to do. (Luqman 31:15)*

And We have enjoined upon man goodness to parents. But if they endeavor to make you associate with Me that of which you have no knowledge, do not obey them. To Me is your return, and I will inform you about what you used to do. (Al-Ankabut 29:8) The verses of both these surahs reveal the correctness of quantum physics in both meaning and depiction. This is because, on the basis of matter-wave fusion, finite human beings experience many things just like finite physics exists in the matter of crossovers. But human beings understand only a little of what they have been through and can never remember it the same way. People have neither ability nor authority because in the case of duality, changes in their own positive or negative values can be realized bringing about infinite possibilities. For this reason, Einstein's saying, *Only the fool needs an order; the genius dominates over chaos.* reveals once again that science determined by people and seemingly realistic is only a tiny part of reality.

4.4.7 THE TRANSFORMATION OF SUBSTANCE

that day turns everything
it touches into something else

today water
has become the sun

today compassion
has become the sun

from wherever it touches
rises love

has the love of knowledge
turned the sun into water

*who has taught it
this knowledge*

*who has gone up to this object
and told it to become water*

The dream of medieval alchemists was to transform the bullet, an element, into gold. The basis of their belief was to make something worthful that was worthless. The first success was achieved by the theoretical physicist Rutherford, who transformed the element thorium into argon. But the world of physics has not moved forward much. In fact, the transformation of *worthless* elements into *valuable* ones means only one thing, which is the transformation of particles that have been left behind during transformation into knowledge. If it were not so, the transformation between elements could not be achieved because rupture and junction particles allow movement and mobility. One of them gravitates towards the bivalency of the soul and the other towards the truth. The fact that both cases are independent and dependent of each other is the starting point. The implications are not inconclusive. On the contrary, there are always principles or methodical concepts that make up the cause of anything. This is because all elements or objects are not just concepts for a solution. In the input and output system (soul), everything carries solution in its own form and has to do so in order to transform matter into knowledge.

Another perspective regarding the transformation of substances is that for the formation of a solution, each substance can be transformed into another substance as a result of the possibility of these infinite conditions. However, the question that needs to be explored here is how this happens. For example, how can any substance be converted into silver? Transformation is possible with correct and regular information regarding the essential and target substance required. This is the information given regarding the elements, structure and functions of atomic elements during the creation of chemical knowledge. It can also be stated that the existence or the nomenclature of quarks in the atomic structure are actually terms used to express a hierarchical atomic function. However, for other forms of functionality, quarks are subject to regulation in their parts according to the principle of electrons, protons and neutrons.

4.4.8 RUTHERFORD AND *THE HALF-LIFE*

The British physicist Rutherford discovered some laws during his work in physics. He named *half-life* one of the things he discovered. What is expressed by this concept is the amount of time spent to convert half of a radioactive element into another element. The best clues regarding these time values of the transformation and the results can be obtained by crossovers. This is because the amount of time required in the crossover process is determined by various parameters, which are the demands of both the speed of visible and invisible matter and magnetic fields. The point to be remembered here is that space already exists. Accordingly, all of these components, with their special functionality, enable the quantum to be constructed in a continuous or indefinite change.

By the way, neutrons and protons take up the task (circuit) of the electrons. In other words, each one can perform the task of the other, from the smallest part to the parts that come together to form its unity. These properties derive from their ability to convert electrons to protons or neutrons. It is even possible to perform the function of the other, i.e., as electrons, to conduct the task of the protons without performing any transformation. This is an outstanding stance, because the most difficult thing is to take on the role of other parts in the same order without changing oneself. For this, infinite knowledge is needed. Feelings in human beings correspond to this infinite knowledge. Crossovers are guided by the pieces on the threshold of this knowledge. Allah does what He says and has it done. This is the basis of truth regarding the mortality of human beings.

4.4.9 LUDWIG WITTGENSTEIN AND THE CHESS OF DUALISM

One of the most difficult mind games is thought to be chess. However, everyone indeed plays this game every day from the perspective of various events. This can be done in family and social life to overcome problems or to win in competitive environments. An interesting feature of the game of chess is that the pieces have been given the right to replace another piece. Ludwig Wittgenstein is the pioneer philosopher and sociologist who likened the discourse of man to chess. According to him, the discourse of man is far from reality. Human beings, who live within the form of duality and its framework, always have a purpose. Their goal is always to reach something. Therefore, he changes his attitudes and behaviors towards these goals. In other words, these behaviors that arise from the psychological building blocks such as the feeling of self-proofing and hiding their fears are always reflected into their language and behaviors. Rather than himself, the person permits his characteristic to take and give shape based on the circumstances descending upon time-space. However, if the person receives letters from the nucleus of atoms, namely, words directly from the mouth where they are located without even touching the events, then they could come out without losing their accuracy. This is possible only through patience and faith in Allah. And there is a force as strong as patience, which is love. Loving the created for the Creator's sake is the respect and love that we find in our sadness and gratitude.

4.4.10 DIRECTING ELECTRONS AND THE KAABA

The only female physicist in the field of quantum physics during the early 20th century, Lise Meitner, found striking results during her research on the element of protactinium. When she examined the beta rays of an element like protactinium, she concluded that the electrons must have a property to enter the nucleus of the atom and to return back from there. The question Meitner shed light on was the fact that beta rays could take all kinds of energy. As a result, electrons are released again with a certain amount or a certain potential. Only in this way, all particles would be able to produce stability according to their needs. Most need to find the direction of electrons outside the nucleus. However, quantum lives have a great role here in providing the necessary ratio for matter and energy. In a sense, if the information about them is understood in the context of truth, so much change in the nucleus would be the case to reduce the energies of electrons outside the nucleus.

In his work on beta rays, the physicist Gamow explained this by throwing particles through the atomic nucleus during radioactive decay. In beta decay, for example, the neutron is divided into a proton and an electron. When the dividend is electron, the electrons go out of the nucleus in a demonstrable way. This is considered impossible in the classical physics approach. However, later on in their research, Gamow and his colleagues found that electrons, rather than penetrating the nucleus wall allow particles to exceed potential obstacles (wall) through a tunnel. Benefiting from many physicists' pre-studies, among them being Rutherford and Meitner, Gamow and the physicist

Ronald W. Gurney and Edward Concon, whom he worked with, named their discovery of the year 1928, which founded the alpha disintegration, *tunnel effect*. Finally, in 1929, the physicist Oskar Klein put an end to the knowledge of this accuracy of tunneling through obstacles via very fast particles.

In fact, are there only electrons or rays passing through the core? More importantly, which electrons and for which purpose are the rays sent out of the nucleus? And what are the sizes of the ones that have to stay in the nucleus? In the Surah Ar-Rahman, Allah revealed the truth of this issue with the verse that He had sent down in the context of man and space: *O company of jinn and mankind, if you are able to pass beyond the regions of the heavens and the earth, then pass. You will not pass except by authority [from Allah].* *(Ar-Rahman 55:33)*

In this sense, is it electrons that explain best the fact that we, as human beings, are defective? Are electrons bivalences? If so, is their main function to create duality? The answer to these questions is that rays emitted from the atomic nucleus (transforming electrons into waves) are able to direct the electrons outside the nucleus. Now that this phenomenon of directing is to direct ambivalence, it is directly linked to how much the artificial and real worlds are experienced. It can also be said that when the orientation is fixed at a certain point, then it becomes constant. This constant road is the person facing qibla, Mecca, Fatiha and/or hereafter. But whether the asymmetrical deviations from these artificial conditions increase, or the path of truth are set up, the artifact is nothing more than the truth. Otherwise, nothing could have happened. Zeilinger, one of the theoretical physicists of this century, was closely interested in this artificial situation.

4.4.11 THE UNIQUE PRINCIPLES OF THE WERNER HEISENBERG AND THE EWG MULTIPLE WORLD INTERPRETATION

The first sip from the glass of the natural sciences makes the man an atheist, but Allah waits for you at the bottom of the glass.

Werner Heisenberg

In the principle of uncertainty, it is obvious that the particles, which make up the atom, undergo a numeric change according to the substance to be replaced and therefore a structural change. In summary, the substances are capable of being transformed in response to the electron need of the atomic particles. For example, when the protons are transformed into electrons or the task of many particles can be undertaken simultaneously and/or successively to create a potential difference. This is the main factor in transforming an object into another. But the German theoretical physicist Heisenberg stresses that the existence of such a fundamental factor depends on the monitoring of the existence without any intervention. Unless a viewer gives his attention to the electrons, they are subject to the principle of uncertainty of Heisenberg in quantum mechanics. That is, it has no features yet. This shows that the electron needs information to activate itself in order to reach the name attributed to it. The stronger this information, the more components it has and the deeper it immense. In momentum, this serves to optimize ratios. When that electron begins to be monitored, then the feature in the purpose of this observation becomes the electron here.

In fact, this attitude of electrons best explains the possibilities that are not recognized according to a quantum particle approach or experienced yet somewhat invisible, and are also the center of the allegory of Schrödinger's cat. There are, therefore, informational components that are invisible besides visible ones. For example, senses are the closest to mathematics and they can describe it best. Although common features of both the visible and invisible constitute the situations, feelings are an attempt of an entirety, i.e., all knowledge, to concentrate on the fancied knowledge.

4.4.12 THE COMMAND OF ALLAH

Indeed, Allah is not timid to present an example - that of a mosquito or what is smaller than it. And those who have believed know that it is the truth from their Lord. But as for those who disbelieve, they say, "What did Allah intend by this as an example?" He misleads many thereby and guides many thereby. And He misleads not except the defiantly disobedient, (Al-Baqara 2:26)

There is no moment in life that is not observed, experienced or felt according to the dimensions and duties of His power. Moments must only become permanent to be active or passive in other parts of the change. If they had not been transformed, the movements and mobility ratios required for dualism could not have been adjusted at all times and the demands of the times could not have been met. Another important point to note is that the reality, i.e., the truth, has its own trajectory and the ones on it, and that it has a duality according to its structure. So to say, the self makes exchanges with the times and makes it possible for the situation to turn into spaces on the basis of the principle of uncertainty regarding movement and mobility. The measurability of the particles in the atom is realized by monitoring them. Nevertheless, whatever the observer feels or thinks, whether clear or ambiguous, i.e., whatever aura has covered the research subject, the measurement data is gained accordingly. Therefore, the knowledge of the researcher shapes the researched and with the formation of the particles, the meaning dimensions are explained according to the ratio of the shape. So far so good, if it is the impression that determines the form of that shape, what is the very thing that enables the observing eyes to have a systematic dimensioning? In the same way, how does writing with a pen happen? What is in the eyes or more precisely, in the senses, there happens a gathering for the formation of waves?

The answer to these questions is hidden in Allah's breathing spirit from His soul. Therefore, the soul is independent of us, but it has incredible knowledge that connects us together. It is clear what the mind or emotions express, even though it is abstract. But is it possible to say the same about the soul? According to the Holy Qur'an, human beings have little knowledge regarding the soul. There is one thing that man has no knowledge about, which is the knowledge of the unseen. The knowledge of the unseen means the formation of space-times from spaces and times. In other words, in the Holy Qur'an, Allah commands the Prophet Muhammad to tell people that he has no knowledge of when the apocalypse will break and that non but Allah has the keys to the knowledge of the unseen. Despite this, human beings, who lack awareness of the knowledge of the unseen, necessarily experience the unseen. In other words, Allah, in the Holy Qur'an, ordered Muhammad to tell people that the Prophet has no knowledge of when the doomsday will come and that the keys to the knowledge of the unseen are owned by none but Allah. Despite this, people who lack the awareness of the knowledge of the unseen will definitely have it. This is because their lifespan progresses. What if one can change the infinite uniqueness of the keys? Can he prevent the accidents or troubles that may happen to him and others?

It can be thought that human beings have an effect on the gates of the unseen left open by Allah. All the sincere worships made in the path of Allah foreshadow good fortune. To be able to comprehend these links in the power of Allah or to feel them from different perspectives requires knowledge. There is the love of Allah in this knowledge. So, the basis that makes this knowledge comprehensible is the command of Allah. Allah's commandment is destiny because it is to be experienced. For example, in the 21st verse of the Surah Yusuf, it is said, *Allah is able to fulfill His command,* and this summarizes the nature of the soul in a spatial context internally and externally. The term internal and external expresses the degree of proximity and distance between the dimensions from a spatial point of view. Seeing that when one observes something, he thinks and perceives what he sees (hence, the things that he perceives become what he knows as space-time and produce a dimensional meaning by intersecting at the linear level), then it is evident that Allah's command is revealed through the sense. In short, is it wrong to say that electrons are the place of Allah's commands to become a word by acting as a means?

According to Einstein and many other scientists, feelings are an entity, i.e., mathematics itself. As the knowledge power of human is limited, it is not enough to interpret this whole as an entity. This situation best describes the first revelations that were sent to the Prophet Muhammad. In fact, there are two verses in this regard. One of them is the Surah Al-Muzzammil, and the other is the Surah Al-Muddathir (74:73, 74). Without these instrumental meanings of the electrons, all the particles in the structure of the atom would not have had the ability to transform into each other.

From another perspective, protons, for example, put electrons in alignment according to their respective angles of interest. The time of alignment determines the rate of tension, i.e., the size of the concept of knowledge. The capacity of protons to align electrons is possible through their understanding of the language. To be able to understand their language is possible only by having their grasp, even being them when necessary.

Neutrons expressing the soul take the good and bad knowledge in the protons and direct their mobility in the orbits of electrons through a signal they leave out. It is the mathematics of an enormous cycle that is explained here. Without these states, there would be no eternal values of truth and variability in dualism. The knowledge, however, is such that it turns back, becomes knowledge, and reaches Allah. In the Qur'an, it is said that Allah is closer to man than his jugular vein, which is hidden in here: *Allah's glance is a single word.* So, Allah is always everywhere. This is because His gaze is knowledge and the human being, which is the book living in the universe. Allah's gaze that is nothing more than a word is His command. In short, everything comes from Allah and returns to Him. Classical physics is sufficient to define the real expectations of events that are not yet known in the space-time continuum. According to Heisenberg, one of the pioneers of quantum physics, everything is prewritten. Events can only be carried out when they are followed. The importance of positioning is that there is a condition or balance of the poles for the effect and reason of existence.

5 A HEART FOR SYMMETRY

5.1 HANDS HOLDING A PEN AND THE HEART

but arrogance
is nothing to us
...
you're lying down on the lawn chair
darkness has scattered into the night

the tree of life
obviously having witnessed
many storms

it listens freely
to all shades of gray
in octaves

silence softly draws
Van Gogh's Starry Night
once again

When does something become comprehension and a product of the imagination? One of the most negative characteristics of man is that he runs after something very fast. He wants what he has been chasing for to happen so quickly or he accepts the imposition of the people around him. Consequently, mental energy is not used in decisions made without thinking. However, the attitudes and behaviors shown in this way affect the way they think. Thus, attitudes and behaviors become an artificial state by creating an opposite projection. As the units of the truth—to be conscious through the mind and to keep consciousness fresh—are almost disabled, the unfortunate consequences of this neglect emerge with the replacement of the mind by attitudes and behaviors. As such, attitudes and behaviors create an artificial mind and shape the task of the mind by giving them a form.

These types of superficial thinking schemes and the resulting thought mechanisms ultimately lead to the growth of human artifacts. These become the coordinates of one's life, so he can persecute his soul by living his life according to these patterns. This artificiality desensitizes him and most importantly, makes it difficult for him to realize his own values. In that, he cannot learn to walk in the orbit of life. Artificiality is the legitimization of the basic values such as religion, science and economy that make up society. However, Allah has given every man a certain potential to realize his own values. The concept of potentiality is much deeper than the concept of intelligence and explains to everyone the best way of being. This is because Allah has created everyone according to a certain measure.

Everyone has their own strengths and weaknesses. These also determine the structure of the person. The best way to explain the concept of potentiality is to ask questions. Benjamin Franklin, the American scientist, philosopher and author of the 18th century, demonstrated this with his aphorism; *The illiterate does not ask questions,* in that the question word is an instrumental method for achieving the degree of wisdom of knowledge within itself. On the other hand, a scientist uses this questioning technique with precision and self-sacrifice to reach a conclusion in his research. He even allocates a large part of his life for research. In such cases, scientists can sometimes be excluded from economic, political and social development, and hence their work cannot develop in the society, which they are a part of. However, if they value the time they cut down from other people in the best way, the octaves that flow into time and from time to people will end up making the same tone.

Naturally, colors, like sounds, are a single tone. When a single tone and color changes into gray tones over time, which is the most important parameter of human life, a sense of responsibility towards the environment falls below the threshold of awareness and indifference increases. This is because the frames of truth, love and respect are experienced at this time. In places where the colors of the heart are limited, one puts a wall between Allah and himself. But those who know Allah are different. People who know Allah the best always live in society out in the open. This is because life has endless colors and jihad in the way of Allah for the beauty of different people who may lead to the completion of the light. Those prophets who know these things surrender themselves to Allah and are worthy of the greetings of Allah being known to having been instrumental in the completion of the light. They are the monuments of the heart and that of the Prophet Muhammad.

Yes, everyone is a book, and everyone is already writing their own book. The books on the shelves in publishing houses tell only parts of the truth as they relate to how the writer sees the truth.

Moreover, the particles of faith, love and truth appear most clearly in knowing Allah and surrendering to Him. In fact, the servant of Allah touches every truth with His love. Nevertheless, an eternal truth awaits him. They are also the books near Allah as well as their own book, which is the account of man's own existence. These are given to Allah completely.

As it is understood, knowledge has an absolute Kaaba, which is the heart. The heart that is created with the essence of the hereafter is antecedent and permanent. Thus, in this world, everything except the heart dies. As the heart doesn't die, the ones in the heart do not either. The more a person remembers the person he loves when that beloved one is detached from life, the more the diseased exists there and he receives sincere love and greetings from the heart. The importance of standing by one's circle, instead of marginalizing it and establishing empathy should be understood better now.

The seriousness of keeping the immortal out of knowledge has covered the history of mankind because in the hands holding a pen, the heart was always missing. The more the heart is missing, the greater the deviations in human life become. Deviations don't help. Moreover, they rapidly increase the artifacts of the world. All the necessary elements of healthy development at its foundation get almost dusty and disappearance of the colors of love becomes inevitable. *The example of those who disbelieve in their Lord is [that] their deeds are like ashes which the wind blows forcefully on a stormy day; they are unable [to keep] from what they earned a [single] thing. That is what is extreme error. (Ibraham 14:18)* Ash is colorless, soulless, mindless, heartless, insensitive, and therefore meaningless. Nevertheless, Allah's eternal mercy bestows the destiny of love, respect and patience to the very destiny drawn by helpless servants. They help them visit the gardens of paradise while they are still alive.

5.1.1 QUANTUM AND THE HEART

For centuries, the mind was dominant in the philosophy of life of the Western world. Up until now, undoubtedly, the throne of the mind has not been shaken. Even mind has been seen as the twin of economic power. From an enlightenment point of view, the human mind is central to everything. It is a research subject of scientists whether there is something that human beings cannot achieve through their minds. This effect continued until the 15th century, when the astronomer Copernicus brought about a milestone. In fact, for many years after him, the idea that objects revolve around man and his unique world had been dominant. Particularly, the religious community has hidden true information from society. But Allah promises that

He will place His word and quantum physics does its part in this.

This physics, which can sustain its own mechanical system with bivalence, is able to maintain the full compatibility and mobility of particle gravity. There is, of course, an increase in the diversity of bivalence. Fast life and fast consumption are negative examples that affect each other. Now, in science, calculations regarding how this increase is augmented or how shortcomings can be eliminated are neglected. The starting point of science is to eliminate the problems that are always present, not to only eliminate them when social problems occur. However, the idea of recovering all areas of humanity from existing problems will be quite utopian. As this is an attempt to force an out-of-order regulation to move to another system, the problems that concern society shall not be solved, and only failures will be repeated. If the function and functionality of quantum mechanics is bivalent, then how can the flower gardens of a dignified life be served to people?

5.1.2 THREE HEARTS AND A MIND

There is always an interaction and communication among the creatures, as is often mentioned here. Therefore, in the context of relativity, movement and mobility are protected by interdependence and support of each other. Allah has given the man a mind and a heart in his creation. Both of these have hearts too. So, walking three times with the heart and once with the mind reduces the margin of error and turns infinite imperfections into goodness. The source of this kind of favor is that people shape their lives, as often mentioned in the Qur'an, with a sound hearth and good sense.

A brief glimpse of this current interstate social order will sufficiently reveal that the heart has been eliminated. This means that the necessary thing for the self-balance of dualism is dualistic structures in reaching the human and humanity itself to make the infinite colors even subtler, instead of gray tones. How does this system work smoother than the other? The parameters of need to fulfill duties directly with the responsibility and leave the work to the epert are clearly explained in the Surah Al-Maa'ooun and At-Takathur.

5.1.3 QUANTUM PHYSICS AND THE SOVEREIGNTY OF THE HEART FOR THE HEART

Finally, quantum physics shows that the sovereignty of reason and mind trying to place the person and his skill in the center is wrong. For centuries, heart and emotions have played a part and a half role in this. Supreme power now reveals the necessity of the awareness of the heart. Otherwise, the rapid disappearance of the universe cannot be avoided.

The meaning of rationalism is based on the belief that it can reach truths through the mind or reason without the need of anything. The history of the representatives of rationalism dates back to Parmenides, Socrates, Plato, Aristotle and Al-Farabi. The most impressive rationalist names in Europe, Descartes, Spinoza, Leibniz, Kant and Hegel, systematized different perspectives on rationalism. Descartes, for example, went from the distinction of body and soul to the idea of *ontological dualism* in metaphysical evaluations. Leibniz gave a different perspective to metaphysics and formed *monads* by placing living and non-living things in the universe in a certain hierarchical order. According to these monads, human beings are above all beings. Kant, in a critical way, approached the existing rationality and shaped the mind in a unique way. Kant was one of the founders of the enlightenment period in Europe.

There have been different currents in the Western world. For example, new Kantianism, which emerged in the 19th century, is a viewpoint, idealism and positivism that continues until today. In the history of Eastern philosophy, or in Islamic thought, the mind has been handled differently. Due to their sense of contemplation and trust in Allah, the science that relates human beings integrates with the knowledge of Allah as knowledge-enlightenment and this never changes in Islamic thought. Although the heart has always been neglected in the Western world, the consciousness of the heart has not been fully realized in the eastern realm. Because of the awareness of the heart, which can be less or more in one than the other, the function of the heart has been neglected. Therefore, it is clear that today it is not the attitude of the West only to consider the mind as the inventor of science.

Violence, which is common in all countries, sheds blood in the world, and in this case, there is a need to develop sensitivity towards the heart urgently. Scientists much responsibility in this. Now it is necessary to break down the idea that the range of the mind will remove various scientific dimensions, such as the heart, against the universal system of science. Allah has already presented it through quantum mechanics. What are such deficiencies? There is no place for emotions in the work of the scientist and in the criteria of his work. However, does not the insufficient function of the heart within an organism accelerate the death of the body when asymmetrical deviations go out of measure?

Human beings are created with the essence of bivalent nature. Therefore, attitudes and behaviors are asymmetric. Allah's asymmetric creation of people is an indication of symmetry. If it weren't already His symmetry, the truth would be unknown. What are overlooked here are the effects of dualism on its functionality. To make things clearer, there is a balance of mind and heart that constitutes this order. This is a commitment relationship of all elements in which the basic ones, such as loyalty relations and existence should be taken into account. The fact that the creatures interact with each other is an expression to each other about themselves, because knowledge is in everyone and everything. Therefore, it concerns all creations and this works in all areas in a synchronic way.

The knowledgeable quantum particles do their part. What about the soul? Allah has given little information about the soul, which can also be called the *secret cube*. The soul provides adaptation and acceptance solutions for the continuous construction of the molecular system in the body. The reason why it provides these analyses is that the soul itself is the knowledge of the essence. So it could be anything. Therefore, it renews itself. With its renewal, it protects its secret and if the Sun could become water for the object, then it can help other things become something else.

5.1.4 QUANTUM, THE HEART AND THE THEORETICAL PHYSICIST WOLFGANG PAULI

One of the pioneers in the defense of the urgency to save science from intellectual monopoly is the Austrian quantum physicist Pauli. According to him, building a perfect system with a very late internal calculation is possible by including the heart in all scientific disciplines. Pauli came to the synthesis of this thought and the disaster caused by the atomic bomb invented by science. It is undeniable that anyone who leads such evil is wise. Using merely reason relies on familiar and simple mathematics. Yet life itself has never been treated according to this calculation. For this reason, human inventions have forced man to be a separate entity from life. In a sense, this scientific approach leads to an increase in the diversity and negation of life, and then it takes advantage of this increase. Today, the multiplicity of taxes applied makes it clear. The lobbying system is a manifestation of the exploitation of mankind by the method of taxation developed by all existing units as a tool.

5.1.5 THE OPERA AND THE LOBBYING SYSTEM

How and in which context does lobbying actually work? The example of the opera clarifies the lobbying system. Every nation has its own opera. The lifestyle of the nation is transferred onto the scene through political, social and current events, costumes, decor and music. There are queues for the audience in front of the opera stage. There are also high, spectacular and high-rise seating areas. Only elite people sit in these areas. They have embraced the idea that they are the privileged people of society. They look at the stage from above to see what's happening. They do not interfere with it, but sit in the upper seating area so as to be considered involved. This is because the subjects as a whole on the stage give a clear impression about how they can serve the understanding of what exploitation is.

There is an understanding of exploitation, which is the result of the feudal system of Western countries, and in which the church has also been a collaborator. France and countries like Britain conducted a policy on exploitation over third world countries such as those in Africa and India. This latest form of postmodern lobbying exploitation has not replaced the elite team, capitalism and imperialism. On the contrary, this system of exploitation has moved forward by developing itself according to time. On top of this system, a hierarchical lobbying exploitation scheme, which concerns religions, states and companies, has been introduced. Rather than imagining the evil of such exploitation that will proceed in the future, it is obvious that the colors and values of the heart must be urgently operated on to save mankind. There is, of course, a solution to everything in life. Turning strange situations into secrets or turning darkness into light should no longer be a dream. Without giving an opportunity for evil, like Gog and Magog, making them remain buried at the bottom of the earth is possible through a world to be established by the heart of quantum physics. The important thing is to realize that there is a need for salvation and also to step up to the plate for humanity.

5.1.6 GOING BEYOND THE ORDINARY BY TAKING ON RESPONSIBILITY

which step of the stairway to the sky
am I on?

before I take steps to the sky
I am confused, counting
as I always look up

one step up
and one glance at the sky

it is so

*what I know and what I don't know
is my meeting with the first step*

*in the dark
and in the light
I always look at the sky*

*and the presence of these steps
makes him love
the glance at himself*

The difference between memorization and interpretation has never been addressed in terms of what they leave to the universe. What is memorization? Memorization is learned even though it is unknown in the first place. Thus, it is crossed over into the same space as the learning is repeated on the same axis with minor differences. There is an illusion in the sum of the values of knowledge. Everything seems planned, but in fact, the heart and mind are attached to a mechanism, so they serve it. This has only one reason: In the universe, everything is *swimming* while ignoring this flow. Denying the flow is to deny the fact that water particles, which make the river a river and the living and inanimate things in it are moving, and to ignore the order.

If real life were like this, life would not have had time because the dead doesn't need to live. In this case, the wind would not be a harbinger of rain nor the seeds would be moved from place to place. However, a careful viewer sees that everything returns and something always changes during the rotation. By ignoring such a life, he will not disappear. However, he lives separate lives from the life that ignores him. Establishing different lives is, of course, possible because time is a variable depending on space and movement. But here is the difference. This person takes a piece from the original place and fills the frame with time. That is, he narrows it down because he runs the limited frame more than enough. On the one hand, the holistic perspective of real life is neglected and on the other hand, the small lives of this life are overcharged. The result is a mechanized life devoid of spirit because of the circles that are affixed on the growing common denominators of these lives. But when a man realizes that he is inadequate and helpless, i.e., if his material-spiritual power is drawn, then he falls painfully from his delusional lives to real life of which he is a part of.

The most determinative factors of individual and social artificial worlds are economic power and position. Thus, how can such memorization be broken down so that a person knows that he has never bathed in the same river and he cares about the knowledge that is the quality of alms? Allah gave people two things, thinking and loving, opening all the doors and affecting all the constituents of the human structure. The knots knitted with these are patiently dissolved one by one. The grain becomes clearer and the treasures of the day are delivered back to this day. Therefore, let our matter be to learn and know this. Let the satisfied man understand the hungry. For this, let us be among those who appreciate the almighty and the righteous deeds so that we become mightily worthy. Only like this can we be so hopeful. Our hope is that Allah and His manifestation emerge within us as Muhammad Mustafa. In everyone and all things, Allah becomes visible. While something new is causing an extreme reaction in us, let us know that it is a manifestation of Allah's command to understand other people who feel and think otherwise. Instead of sending those who have reached the knowledge of *Anal Haq* like Mansur Al-Hallaj to the gallows of knowledge, we should get a benefit from the great knowledge that they have, that's the thing.

5.1.7 THE DOMINO EFFECT AND JUSTICE

Let's assume that the life we live is a road map made up from a set of dominoes, as in the experiment of Schrödinger's cat. One of the dominoes at the very beginning of this array must be knocked over by someone, so that the others can also fall down. The person who knocks over this domino can influence the differentiation or alteration of a situation. Endless possibilities can be put forward here. Examples of these possibilities are that the other dominoes could fall over quickly, slowly, successively, or to the right. If the person chooses not to do anything, the dominoes remain in their lined up form. They may fall down again for any reason this time other than that person. This information is given to the dominoes (i.e., atoms). This is because possibilities can happen as destiny thanks to the principle that something nonexisting cannot happen. Another situation is that it is unknown who or what touches the domino. In this case, only the state of the dominoes being knocked over is apparent.

Another example takes into account the infinity of possibilities by explaining that all possibilities are absolutely manifested in different dimensions both in an individual and in all persons, beforehand and eternally, as well as simultaneously. It is important to remember that this example is only a road map. Everything created in the universe hsa to leave leave its trace or work completely to all its road maps.

In summary, it is quantum mechanics, which takes over the crown of classical physics, and which is inevitable in every aspect of our life, basically carrying the visible and invisible in all its accounts, and which cannot be refuted. Therefore, why hasn't this science, which has been a part of our environment for over a hundred years, and which tells the truth, shed light on our road maps? In other words, why isn't it used in the justice system? If we want to take the justice of

Allah in our intentions, then it is our duty to show them in our behaviors.

It must be accepted that the distortion of memorization is now inevitable. Justice must listen to the beat of the heart. The mistake of the human being can be compensated in a place where there is the heart. The heart is creativity. Wherever there is creativity, there is Allah, and He shows the truth to man. Otherwise, the innocent citizen is afraid of justice as well as being inclined to commit a crime by being pushed into invisible and systematic crime elements. The trigger for the crime is the spread of phantoms into all areas from the justice of an uncontrolled country.

Since the purpose of the phantoms is to drive people into a corner, they appear preferably in the dark. This is because they know that the darkness is the unknown for people and no one can predict when, where and what they can do. This is their conception of domination formed based on their own worship patterns in the darkness. If we were to conclude with the words of Allah: *Unquestionably, it is they who are the corrupters, but they perceive [it] not. (Al-Baqara 2:12)*

SOURCE:

The Qur'an – Various Interpretations
Muhyiddin İbn Arabî: "Füsusu'l-Hikem"
Mevlânâ Celâleddîn-i Rûmî: "Fihi Ma-fih"
Ahmet Koç: "İhvani Safa'nin Eğitim Felsefesi"
Enver Uysal: "İhvani Safa Felsefesinde Tanri ve Âlem"
İmam-i Gazalî: "Kimya-yi Saadet"
Heidegger: "Brief über den Humanismus"
Heidegger: "Sein und Zeit"
Ian Almond: "A Comparative Study of Derrida and Ibn 'Arabi"
Noam Chomsky: "Human Nature: Justice versus Power"
John Gribbin: "Auf der Suche nach Schrödingers Katze"
John D. Barrow: "Das Buch der Universen"
Ernst Peter Fischer: "Die Hintertreppe zum Quantensprung"
Dirk Schneider: "Jesus Christus der Quantenphysiker"
Stephen Hawking: "Schwarze Löcher gibt es nicht"
Stephen Hawking und Ron Miller: "Eine kurze Geschichte der Zeit"
Leonard Süsskind: "The black hole war"
Andreas Müller: "Raum und Zeit"
Hans Dieter Zeh: "Physik ohne Realitaet: Tiefsinn oder Wahnsinn"
Matthias Matting: Die neue Biographie des Universums"
Uwe Fahrenholz: "Allgemeine Transformationslehre"

Günter von Hummel: "Nach Lacan"
Pavel Florensky: "Reverse Perspective"

www.ingramcontent.com/pod-product-compliance
Lightning Source LLC
Chambersburg PA
CBHW031603210526
45464CB00004B/1410